ROUTLEDGE LIBRARY EDITIONS:
SOCIAL AND CULTURAL GEOGRAPHY

Volume 1

THE POWER OF PLACE

THE POWER OF PLACE

Bringing together geographical and
sociological imaginations

Edited by
JOHN A. AGNEW
AND JAMES S. DUNCAN

Routledge
Taylor & Francis Group

LONDON AND NEW YORK

First published in 1989

This edition first published in 2014
by Routledge
2 Park Square, Milton Park, Abingdon, Oxfordshire OX14 4RN

and by Routledge
711 Third Avenue, New York, NY 10017

First issued in paperback 2015

Routledge is an imprint of the Taylor & Francis Group, an informa business

British Library Cataloguing in Publication Data
A catalogue record for this book is available from the British Library

ISBN: 978-0-415-83447-6 (Set)
ISBN: 978-1-138-99804-9 (pbk)
ISBN: 978-0-415-73320-5 (hbk) (Volume 1)

Publisher's Note
The publisher has gone to great lengths to ensure the quality of this reprint but points out that some imperfections in the original copies may be apparent.

Disclaimer
The publisher has made every effort to trace copyright holders and would welcome correspondence from those they have been unable to trace.

THE POWER OF PLACE

Bringing together geographical and sociological imaginations

EDITED BY
John A. Agnew
James S. Duncan

Boston
UNWIN HYMAN
London   Sydney   Wellington

© John A. Agnew & James S. Duncan, 1989

This book is copyright under the Berne Convention. No reproduction without permission. All rights reserved.

Unwin Hyman, Inc.
8 Winchester Place, Winchester, Mass. 01890, USA

Published by the Academic Division of
Unwin Hyman Ltd
15/17 Broadwick Street, London W1V 1FP, UK

Allen & Unwin (Australia) Ltd
8 Napier Street, North Sydney, NSW 2060, Australia

Allen & Unwin (New Zealand) Ltd in association with the
Port Nicholson Press Ltd,
75 Ghuznee Street, Wellington 1, New Zealand

First published 1989

British Library Cataloguing in Publication Data

The power of place : bringing together geographical and sociological imaginations.
1. Social conditions. Geographical aspects
I. Agnew, John A. Duncan, James S.
304.2
ISBN 0-04-445281-0

Library of Congress Cataloging-in-Publication Data

The power of place : bringing together geographical and sociological imaginations / John A. Agnew & James S. Duncan, editors.
    p. cm.
Essays based on lectures given at Syracuse University, fall semester, 1986.
Bibliography: p.
Includes index.
ISBN 0-04-445281-0 (alk. paper)
1. Anthropo-geography. 2. Social sciences. 3. Geopolitics.
I. Agnew, John A. II. Duncan, James S. (James Stuart), 1945–
GF49.P68   1989
304.2—dc19   88-27683
                                                         CIP

Typeset in 10 on 11 point Bembo
Printed in Great Britain at the
University Press, Cambridge

Contents

Dedicated to the memory of David E. Sopher
(1924–1984)

Preface

This collection of essays is the fruit of a series of lectures given at Syracuse University in the fall semester of 1986. One purpose of the lecture series was to examine some of the uses of the concept of place in social science and history, especially in relation to questions of power and politics. Another purpose was to examine why recourse to the concept of place has been both infrequent and idiosyncratic within the confines of normal social science and history. Some essays address both purposes, if with differing degrees of emphasis. Others tend to focus on one rather than the other.

The lecture series was organized and this book is published in memory of David E. Sopher, who died in March 1984. David Sopher was a major influence upon a number of students and faculty colleagues at Syracuse University who were privileged to know and work with him. Broadly educated in a time that values narrow technical expertise, David Sopher cherished a real world of cultural variety and complexity over what he believed to be an *intellectual* world of universal motivations and simple laws of social behavior. It was in this context that the concept of place loomed large for him. Places provided both the real, concrete settings from which cultures emanated to enmesh people in webs of activities and meanings and the physical expression of those cultures in the form of landscapes.

By no means all of the contributors to this book knew or worked with David Sopher. In this respect it is not a typical memorial volume. But they all share a common interest in exploring and illustrating the history and uses of the concept of place. In this respect they are David Sopher's intellectual heirs at a time when there is a general revival of interest in place and places at the expense of the universalizing or law-seeking social science against which David Sopher was often embattled. We only wish he were here to join in the fray.

John A. Agnew
James S. Duncan

1

Introduction

JOHN A. AGNEW & JAMES S. DUNCAN

This book is an attempt to make the case for the intellectual importance of geographical place in the practice of social science and history. It reflects the recent revival of interest in a social theory that takes place and space seriously. Foucault (1980, p. 149) captures what is most at stake when he writes:

> Space was treated as the dead, the fixed, the undialectical, the immobile. Time, on the other hand, was richness, fecundity, life, dialectic.... The use of spatial terms seems to have the air of anti-history. If one started to talk in terms of space that meant that one was hostile to time. It meant, as the fools say, that one "denied history".... They didn't understand that [these spatial terms]...meant the throwing into relief of processes – historical ones, needless to say – of power.

Place is a difficult word. *The Oxford English Dictionary* gives over three and one-half pages to it. It can mean "a portion of space in which people dwell together", but it can also mean "rank" in a list ("in the first place"), temporal ordering ("took place"), or "position" in a social order ("knowing your place"). In modern social science the geographical meaning has been largely eclipsed by the others. In particular, the social categories of state censuses have pre-empted geographical places as the major operational units of social theory; classes and status groups have displaced places and geographical settings. In sum, sociology has transcended human geography.

The last few years have seen a surge of interest in the possibility and importance of bringing together what can be called the geographical and sociological "imaginations". The geographical imagination is a concrete and descriptive one, concerned with determining the nature of and classifying places and the links between them. The sociological imagination aspires to the explanation of human behavior and activities in terms of social process abstractly and, often, nationally construed. Since the early part of this century these imaginations have been separated by institutionalized methodological and conceptual differences and competing objectives. Recent work in social theory, however, represents a movement towards a meeting, if not a marriage, between these imaginations (see e.g., Giddens 1979, 1981, Foucault 1980).

To date, these developments in social theory have not really filtered down into the practice of history, social science, or geography. One purpose of this book is to suggest some ways in which the two imaginations can be simultaneously engaged by means of a focus on the concept of place. Another is to explore why and how the concept of place has been marginalized within the discourses of modern social science and history.

Approaches to defining a geographical concept of place have tended to stress one or another of three elements rather than their complementarity. Firstly, economists and economic geographers have emphasized *location*, or *space* sui generis, the spatial distribution of social and economic activities resulting from between-place factor cost and market price differentials. Secondly, microsociologists and humanistic geographers have concerned themselves with *locale*, the settings for everyday routine social interaction provided in a place. Thirdly, anthropologists and cultural geographers have shown interest in the *sense of place* or identification with a place engendered by living in it. Rarely have the three aspects been seen as complementary *dimensions* of place. Rather they have been viewed as mutually incompatible or competing definitions of place. But they are related. If locale is the most central element of place sociologically, it must be grounded geographically. Local social worlds (locale) cannot be completely understood apart from the macro-order of location and the territorial identity of sense of place (see Agnew 1987).

The fragmentation of the meaning of place has occurred in a particular historical–intellectual setting. The modernization theories that have dominated recent social science and history have focused heavily on the *national scale*. This has led to a neglect of *scales* other than the national in the analysis of social process. Concepts of place have survived but only in various marginalized and subordinate discourses that have selected the element most appropriate to their needs (Sack 1980).

The recent revival of interest in place does not involve rescuing, restating, or elaborating the various partial perspectives that have prevailed previously. Although the ways in which the elements of place are brought together will differ, as is apparent from the essays in this volume, the focus now is upon engaging them all rather than championing one over the others.

A critical question concerns: why now? Why is it that there is only now an energetic attempt to bring together the geographical and sociological imaginations by means of a focus on a multi-dimensional concept of place? A complete answer to this is probably impossible. We can only suggest some parts of an explanation here. One part of an explanation lies in the crisis that has afflicted *national*-oriented history and social science over the past 20 years. Modernization theories have come in for heavy criticism from a number of directions. Their functionalism, positivism, and evolutionism have been decried on both broadly philosophical and more specific empirical grounds. However, alternatives have been hard to come by. In a real sense there has not been a complete alternative to, only a critique of, the prevailing sociological imagination.

But this intellectual crisis is only part of a wider political and social crisis: the crisis of the modern territorial state and its legitimating myths.

This has a number of sources. One is the fact that the power of the state has been fatally compromised by the emergence of a global economy that is no longer under the control of individual governments, either singly or collectively. Another is that the power of the state itself was never created only by coercion from the top down. Rather, the power of the state rested on thousands of "bits" of minute or local consensus. Increasingly, national consensus has been challenged by pressure groups (trade unions, political terrorists, etc.) which have revealed through their actions the *fragility* of state power and its basis in the *local* rules of social cohabitation.

Parenthetically, it is well worth noting that nation-building itself rested upon the privileging of certain places as capitals, as seats of festivals, or what have been called *"les lieux de memoire"* (places of memory). Even nationalization, therefore, epistemologically so antithetical to place (it is no coincidence that *"lieux"* means *"pot à pisser"* as well as "places" in French) gave to place an ontological variety of new meanings (Nora 1984).

Similarly paradoxical, the growth of mass communications has also reduced the *social* basis to state power. The emergence of mass communications would seem to lead inexorably to mass culture. However, the effect has often been the opposite. The reception of messages depends on interpretation, and interpretation depends on the nature of the sociological situations in which different frames of reference operate. The same stimulus need not generate the same responses. One way of putting this in a political context is as follows:

> to the extent that the salient political cleavages in a society follow, rather than cut across geographical divisions, a national political environment and national media focusing attention on the same issues everywhere may serve to accomplish the opposite of nationalization, that is to stimulate dissimilar behavior on the part of geographical units (Claggett et al. 1984, p. 90).

One particularly important feature of the present period that correlates with the crisis of the state and its social science is the collapse of what can be called the *Pax Americana*. The self-confidence of the modernization theories rested initially and finally upon the perceived achievements of modern America. Insecurity is the new key word, in America and elsewhere. Umberto Eco (1986, p. 79) finds in present-day anxieties a parallel with the Middle Ages. A "clan spirit" has returned:

> But insecurity is not only "historical" it is psychological, it is one with the man-landscape/man-society relationship. In the Middle Ages a wanderer in the woods at night saw them peopled with maleficent presences; one did not lightly venture beyond the town; men went armed. This condition is close to that of the white middle-class inhabitant of New York, who doesn't set foot in Central Park after five in the afternoon or who makes sure not to get off the subway in Harlem by mistake, nor does he take the subway alone after midnight (or even before, in the case of women).

Taken together the varied sources of a "denationalization" of history and social science provide an opening for consideration of a multidimensional concept of place as a means of integrating the sociological and geographical imaginations. In the face of the dissolution of the "national" it is no longer possible for the sociological imagination to ignore the geographical. Or as Eco put it, with an appropriate double-entendre, "let's give back to the spatial and the visual the place they deserve in the history of political and social relations" (1986, p. 215).

The chapters in this collection work around and upon the two major themes identified earlier: the intellectual history of concepts of place and the interpellation of power and place. They do so in different ways. Yet these themes must not be seen, as some detractors of the concept of place might argue, as fundamentally different; the first, abstract, general, sociological, and placeless; the second concrete, specific, geographical, and place-laden. For the intellectual histories, as much as the studies of specific places, are place-laden. They are not abstract intellectual histories, but histories of how, for example, American social scientists have thought about places, how French geographers and historians have thought about places, or how European and American historians have thought about Renaissance Italian cities. The organization of the chapters of this book is broadly geographical. The first four studies deal with the power of place (intellectually and empirically) in North America, the second three with Europe, the next two with Latin America, and the final two with Asia. This organization emphasizes the point that the study of place remains relevant everywhere around the globe. It is not simply a concept which pertains in the "*Third World*" or in Europe and America in the distant past. But we are not engaged in the contemplation of the idiographic. It would be mistaken to read the book this way. We are concerned with the role of place in social organization and social thought, not as an end in itself.

Although all of the authors are keen to stress the salience of place, they approach it in different ways. Some of the authors (Agnew, Entrikin, Berdoulay) provide us with broad overviews of how the social sciences in particular socio-political contexts have treated not only the concept of place but also actual empirical places. Other authors (Muir & Weissman, Richardson) discuss the disciplinary history of the treatment of place while focusing upon one or two empirical case studies. Still others (Ley, Hugill, Cosgrove, Robinson, the Samuels, Duncan) devote their attention to empirical case studies of the relationship between power and place. We feel that these different approaches are equally valid ways of engaging the power of place and understanding its meaning and significance.

The first section of the book deals with place in North America. Of the four chapters in this section, the first two focus upon intellectual history while the latter two provide case studies of the explanatory power of place. In the first chapter political geographer John Agnew examines the devaluation of place in social science. This devaluation is traced in conventional social science to the association of place with "community". With the presumed eclipse of community, place has been eclipsed too. In Marxist social science there has been a tendency to absolutize the power of commodification. The universalization of capitalism thus undermines the social significance of place. However, rather than disappearing in the face of the eclipse of community or the logic of liberal capitalism the *nature* of place has changed. Indeed, commodification and resistance to it generate new pressures for place. Capitalism, therefore, while transforming society, has created a new structuring role for place.

J. Nicholas Entrikin, historian of geographic thought and urban geographer, concerns himself with the bases for judgment about the significance of place and regional studies and how these have changed as beliefs about modernity and rationality have changed. The question of significance, which has a dual sense of both importance and meaning, is divided into three distinct but related parts: empirical–theoretical significance, normative significance, and scientific significance. The first addresses the question of the areal variation of economy, society, and culture; the second addresses the cultural value associated with this areal variation; and the third more specifically addresses the "scientific" status of place and regional studies. All these bases for judgment about significance are related to ideas about modernity in American culture and social thought.

Cultural geographer David Ley traces the relations of space and place to the discourse of modernity through an examination of the "struggle" over the definition and meaning of the built environment. The suppression of local context and culture and the imposition of uniformity as a means to universality are seen as common to both modernist architecture and modernist geography. The post-modern struggle for place is viewed as part of a wider struggle over the definition of culture and attempts to "rehumanize" urban space.

Cultural geographer Peter Hugill examines home and class among an American landed elite in the 19th century. An American elite in upstate New York before the Civil War was a true landed elite on the English model with domestic roots deep in country life and a commitment to the local and the "natural". It was based in class differentiation but one sensitive to its links with other groups and the overall ecology of place. After the Civil War these roots were torn up and replaced by the pursuit of pleasure in picturesque artifacts and the importation of behaviors conceived elsewhere with little concordance to the local. It marked the triumph of social differentiation at the price of lost sensitivity to class differences and ecological complexity. Sturdy farmers were transformed into peasants and robber barons were elevated to nobility. Place was transformed in train.

The next three chapters deal with the power of place in Europe. Social historians Edward Muir & Ronald Weissman explore the social and symbolic places of Renaissance Venice and Florence. They pose the question: what sense have historians made of urban geography in the Italian Renaissance? Their answer is that whereas an older historiography shared many of the assumptions of modernization theories, the recent historiography of Renaissance Venice and Florence has given special emphasis to the sociology of space and geographically-based social ties. A sensitivity to place, they argue, offers historians a rich context for the analysis of

Renaissance society, and one which challenges many common assumptions about the Renaissance social order.

The historical geographer Denis Cosgrove demonstrates just such a sensitivity to place in his study of power and place in the Venetian territories. Cosgrove uses the landscape frescoes of the Villa Godi at Lonedo di Lugo to construct an account of power and place in the 16th-century Venetian landscape. He shows how the flux and uncertainty of everyday life are transformed in the frescoes into an immutable image of harmony. The frescoes, the villa, and the landscape in which it was set are equal participants in the cultural process of creating human meaning and articulating power within the social and environmental context of the Venetian *terra firma* in the mid-16th century.

In the final chapter of this section cultural geographer Vincent Berdoulay assesses the treatment of the concept of place by Francophone geographers and historians. He first discusses the Vidalian conception of place, arguing that it had a major impact upon the *Annales* school of history. The author then examines more recent Francophone work on place, focusing first upon structuralist and in particular semiotic perspectives on the structure of place and subsequently upon those works which focus upon meaning. He then explores recent work which attempts to mediate between structure and meaning by adopting a model of emergence which views places as capable of generating both their own structure and meaning. In the final section of the chapter the author discusses different types of discourse on place and asks what the relationship is between the discourses and the places themselves.

The next two chapters deal with place in Latin America. Miles Richardson, an anthropologist, examines the Germanic intellectual roots of cultural anthropology and cultural geography and their pursuit of answers to the question: where do humans stand in nature's scheme? He explores the use of the concepts "place" and "culture" in the two disciplines both in terms of intellectual history and by reference to the images of Christ found among Southern Baptists and Spanish American Catholics. Baptist and Catholic create specific Christs to be within two distinct places. In Spanish America, the place is the *iglesia*-temple, a place in itself made holy through ritual. In the American South the place is the church-auditorium, where the preacher and the Holy Word replace the ritual. In both places, however, the creative process is to deny death and the limits imposed by earthly places and thus to put death in its place. Man's place in nature, therefore, is to put death in its place.

Historical geographer David Robinson demonstrates that the language of place in Latin America is both complex and rich. The variety of terms in Spanish, Portuguese, and the Amerindian languages allows for an understanding of the social behavior and place imagery of the various cultural groups. Examining the mutation and elaboration of place terms over time shows the persisting significance of place even as its meanings and operational definitions have changed.

The final two chapters in the book deal with place in Asia. Cultural geographer James Duncan examines the politics of place in Kandy, Sri

6

Lanka between 1780 and 1980. Place is defined here as a specific locality that is characterized by a number of landscapes which have been created and modified over time by political and social change under three different cultural paradigms: the Kandy of the god-kings in the late 18th century; the Kandy of British colonialism in the 19th century; and the Kandy of Buddhist nationalist politicians in the mid–late 20th century. The meanings of place and its landscapes have changed during these periods, yet a central narrative structure links the earlier with the most recent ones.

Cultural geographers Marwyn & Carmencita Samuels study Beijing as an example of the salience of place in modern China. Having served as the "celestial capital" of Imperial China, the formal *axis mundi* of the imperial cult, the secular metropole of the Confucian world for more than 600 years, modern Beijing has few equals in terms of the sheer numbers of historic sites, buildings, and artifacts reminiscent of imperial traditions. Its landscape is everywhere littered with reminders of past greatness. However, since 1949 the city has also served as the central repository of socialist China, the administrative core and symbolic hub of revolution. The question that they pose is: how has the traditional cosmo-magical or "celestial" layout and design of the imperial capital fared in the context of 20th-century revolution?

As this overview of the chapters shows, the concept of place elaborated in this book entails a synthesis of both geographical and sociological imaginations. The authors engage in this task in two different but related ways. First, a number of authors show how different disciplines in different sociological–intellectual settings have produced different discourses about place. Some authors, for example Ley and Berdoulay, argue that at times discourses have reflected the nature of empirical places while at other times, as Agnew suggests, a dominant discourse is silent about the importance of place even in the face of evidence to the contrary. One of the major purposes of this volume as a whole is to argue against the prevalent tendency in history and social science to overvalue the sociological imagination at the expense of the geographical one. Second, some authors demonstrate through empirical study how place serves not only as an indicator but as a source of social and political order. They argue that place, both in the past and in the present, both in the third world and in the first, serves as a constantly re-energized repository of socially and politically relevant traditions and identity which serves to mediate between the everyday lives of individuals on the one hand, and the national and supra-national institutions which constrain and enable those lives, on the other.

References

Agnew, J. 1987. *Place and politics: The geographical mediation of state and society.* London: Allen & Unwin.

Claggett, W., W. Flanigan & N. Zingale. 1984. Nationalization of the American electorate. *American Political Science Review,* **78**, pp. 77–91.

Eco, U. 1986. *Travels in hyperreality.* San Diego: Harcourt Brace Jovanovich.

Foucault, M. 1980. *Power/knowledge*. Brighton: Harvester.
Giddens, A. 1979. *Central problems in social theory*. London: Macmillan.
Giddens, A. 1981. *A contemporary critique of historical materialism*. Berkeley: University of California Press.
Nora, P. 1984. *Les lieux de memoire*. 3 vols. Paris: Gallimard.
Sack, R. D. 1980. *Conceptions of space in social thought: a geographic perspective*. Minneapolis: University of Minnesota Press.

2

The devaluation of place
in social science

JOHN A. AGNEW

This chapter involves the merging of two interests. One is an interest in the recent revival by geographers of a concept of place. Many geographers of diverse intellectual and ideological persuasions are talking about place and the need for a revitalized regional geography. The second is an interest in why social science in general and geography in particular should have devalued place to the extent that it must be defined and refined rather than have been seen all along as a vital social science concept.

I want to be clear at the outset that most recent arguments for a place-based geography are not calls for the revival of regional geography as it has been practiced. I have in mind here the regional geography based on either the immutable physical landscapes of fixed regions at a scale intermediate between locality and national unit or descriptive inventories of regional characteristics regarded as if they are independent of social order. The conception of place involved in these approaches could not be expected to excite much interest in social science. Rather, as, for example, in Doreen Massey's book *Spatial divisions of labor* (1984), the call is for a concept of place in which spatially extensive fields of economic and political power are mediated through historically defined conjunctures of social interaction specific to localities. The call therefore is not for geography to abandon social science in a return to an old-style regional geography with regions as arbitrary data collection units but for social science to incorporate into its core lexicon a geographical concept of place.

The focus of this chapter is on why we should have taken so long to arrive at a *theoretically coherent* concept of place as a defining element of geography and key idea for social science as a whole. I should add that it is not my purpose to do so here, though I have attempted this elsewhere (Agnew 1987). A short discussion of the defining elements of place is provided in the Introduction to this book. The major argument is that within much social science two other concepts, those of community and class, have dominated to the extent that thinking and talking in terms of place has been largely impossible. I do not intend to scotch either of these other concepts; indeed each bears some role in recent discussions over the concept of place. My point is that their centrality in the discourses of social science has left little scope for a concept of place.

9

One can identify two major types of social science. The first, "orthodox" or conventional social science, claims to take the world as it finds it. The objective is to find the structural and/or historical foundations of present-day social organization. The second, "antithetical" social science, views the orthodoxy as ideological, functioning to legitimize present-day social organization rather than expose its "real" roots. The most important system of antithetical social science has been Marxian political economy. I argue that both types of social science have devalued the concept of place but that they have done so in different ways. Neither orthodox nor antithetical social science has had much time for place, but for different reasons.

The chapter takes the following form: first, two logical stages to the devaluation of place in orthodox social science are identified: the first involves the confusion of a geographical concept of place with the sociological concept of community; the second involves inferring from the supposed eclipse of community in modern society the eclipse of place. Second, an explanation for this devaluation of place in modern social science is outlined. Third, I argue that antithetical social science in the form of Marxian political economy has also devalued place by accepting the logic of liberal capitalism it purports to critique. But, recently, writers from this tradition have provided much of the inspiration for new conceptions of place. I suggest that this reflects attempts to bring together the manifest reality of geographical variation with a *logic* of historical change that allows for *new* definitions of place over time and space.

The devaluation of place in orthodox social science

Confusing place and community is widespread in the social sciences. This reflects an underlying ambiguity in the language of community as it has developed in the social sciences. Much has been made of the language of class as a fundamental discourse in social theory. But at least as widespread today, and perhaps even more important in the 19th century, was the language of community. This language had an old heritage but it was during the massive social changes of the nineteenth century that it came to have two historically specific connotations that have persisted to the present day. Specifically, community was conceived as both a physical setting for social relations (place) and a morally valued way of life (Calhoun 1980, p. 107). Thus

> The language of community grew up as a demand for more moral relations among people as well as a descriptive category. . . . Community was far more than a mere place or population (Calhoun 1980, p. 106).

Nisbet (1966, p. 18) sees the moral usage of community as historically prior to that of social relations in a discrete geographical setting or place. He notes:

Community begins as a moral value; only gradually does the secularization of this concept become apparent in sociological thought in the nineteenth century (Nisbet 1966, p. 18).

But even while sensitive to competing connotations Nisbet maintains the fusion of the "experiential quality of community" and the *actual* social relationships of place that has come to bedevil contemporary social science. Nisbet (1966, p. 48) argues strongly to the effect that:

> By community I mean something that goes far beyond mere local community. The word, as we find it in much nineteenth- and twentieth-century thought encompasses all forms of relationships which are characterized by a high degree of personal intimacy, emotional depth, moral commitment, social cohesion, and continuity in time. Community is founded on man conceived in his wholeness rather than in one or another of his roles, taken separately, that he may hold in a social order. It draws its psychological strength from levels of motivation deeper than those of mere volition or interest, and it achieves its fulfillment in a submergence of individual will that is not possible in unions of mere convenience or rational assent.

Of course, this viewpoint is not merely Nisbet's but draws on a long intellectual tradition.[1] In Tönnies (1887) *Gemeinschaft/Gesellschaft* dichotomy community had a similarly moral affirmation in the face of the demands of rational and thus non-communal and non-local associative relationships such as social class. To Durkheim the functions once performed by "common ideas and sentiments" in *Gemeinschaft* society were performed in large-scale industrial societies by completely new social institutions and relations. This transformation involved major changes in the nature of both morality and social solidarity. Durkheim's central thesis was that "the division of labor," by which he primarily meant occupational specialization, "is more and more filling the role that was once filled by the *conscience commune*; it is this that mainly holds together social aggregates of the more advanced type" (Durkheim 1933, p. 173). The division of labor was "the sole process which enables the necessities of social cohesion to be reconciled with the principle of individuation" (Durkheim 1904, p. 185). Weber (1925) took much the same view as Tönnies and Durkheim, seeing local community as based on subjective feeling in contrast to the rationality of national social relationships.

The founders of academic sociology in the United States were particularly interested in how the growth of cities affected the constitution of social relationships. Their "programmatic question," as Fischer (1975, pp. 67–68) has put it, reflected that of their European precursors: "How can the moral order of society be maintained and the integration of its members achieved within a highly differentiated and technological social structure?" It was "the problem of community in the New Age."

The typological approach of Tönnies seemed to offer an especially powerful insight into the "urban transformation" through which they were living (Faris 1967, pp. 43–48, Shils 1970, Cahnman 1977). Robert Park, for example, one of the leading figures in the sociology department at the University of Chicago and a founder of urban sociology, noted that though the diverse terms used to label the community – society dichotomy revealed its "unrefined nature":

> The differences are not important. What is important is that these different men looking at the phenomenon from quite different points of view have all fallen upon the same distinction. That indicates at least that the distinction is a fundamental one (quoted in Zimmerman 1938, p. 81).

The reliance on the community–society dichotomy reached its peak in the urban theory of Wirth (1938) with its folk–urban polarity, and the modernization theory of Parsons (1951), with its traditional–modern dichotomy. Tönnies' idea that the modern world was the product of a transition from a place-based community to a placeless or national society was thus perpetuated in the dominant strands of American sociological theory.

In contemporary political sociology, for example, one finds a merging of place with community identical to that of earlier sociology. Hays (1967, p. 154) provides an explicit example:

> One conceives of community as human participation in networks of primary, interpersonal relationships within a limited geographical context. At this level there is concern with the intimate and personal movement within a limited range of social contacts, and preoccupation with affairs arising from daily personal life. Knowledge is acquired and action carried out through personal experience and personal relationships.

But society is national and involves "secondary contacts over wide geographical areas, considerable mobility and a high degree of ideological mobility" (Hays 1967, p. 154). Society is therefore distinct and counter to community-in-place. When the "modernizing" forces of society overpower the "traditional" forces of community, place is overpowered too and continues to exist only as the location of nationally-defined social activities. That modern social activities are or even could be defined in places has become a contradiction in terms.

A few attempts have been made to distinguish place from community and argue for their non-complementarity. For example, Webber (1964) has proposed spatial accessibility (ease of access) rather than the propinquity of place (co-residence) as the necessary condition for community. As Webber (1964, pp. 108–9) says:

> The idea of community . . . has been tied to the idea of place. Although other conditions are associated with the community – including "sense of belonging", a body of shared values, a system of social organization, and interdependency – spatial proximity continues

12

to be considered a necessary condition. But it is now becoming apparent that it is the accessibility rather than the propinquity aspect of "place" that is the necessary condition. As accessibility becomes further freed from propinquity, cohabitation of a territorial place – whether it is a neighborhood, a suburb, a metropolis, a region, or a nation is becoming less important to the maintenance of social communities.

The problem with this is that it resolves the dualism of community in favor of the way of life or solidarity element and at the expense of social relations structured in place, despite tremendous evidence to the contrary. It also fails to account for the fact that even in big cities in "modern" societies people act together on occasion on the basis of common territory (Bell & Newby 1971, Tilly 1973, p. 212). Many communities still rest on propinquity.

At the crux of the confusion over place and community is the ambiguous legacy of the term community. Rather than distinguishing its two connotations, a morally valued way of life and the constituting of social relations in a discrete geographical setting, they usually have been conflated. In particular, a specific set of social relations, those of a morally valued way of life, have transcended the generic sense of community as place. Such an emphasis has, as Calhoun (1978, p. 369) reminds us, discounted "the importance of the social bonds and political mechanisms which hold communities together and make them work." It has also reduced the possibility of seeing society-in-place as an alternative to community-in-place. For orthodox sociology, and social science in general, place and society, and consequently geography and sociology, have been antithetical.

The dichotomy of "community and society" has been a major framework into which social scientists have set their discussions of human association and social change.[2] The theorists whose writings have provided the orientations and terms of discourse for much contemporary social science lived and wrote during the last half of the 19th and early part of the 20th centuries. This was a period of immense social change and perceived disorder in social relations. The American and French Revolutions had set the stage with their new doctrines of citizenship, equality, and individual rights. The Industrial Revolution was creating in Europe and America a new order of human institutions and social classes. Populations had grown exponentially. Nationalism had appeared as a principle of abstract solidarity. New states emerged to change the political map of Europe. It was an era of economic discontent and political revolt. Self-consciously, intellectual elites of the 19th century were aware that what they had been brought up with was now passé; that a new world of human relationships was coming into existence. It was this experienced and perceived change in history that occupied the "social" thinkers of the period (Gusfield 1975, p. 3).

In the circumstances of the late 19th century it is understandable that a major theme of many writers would be the *direction* of history. In particular, a common belief in social evolution led many influential writers to view their time as moving the world along a line of direction away from one point and toward another. The movement involved was from one type of human association to another. They saw each as constituting a systematic arrangement of elements and each as mutually exclusive from the other. The forms of the "modern" broke sharply with the dominant aspects of pre-19th-century life. The modern was both more individuated and individualistic in social organization than the "traditional". Bonds of group loyalty and affective ties had given way to the rationalistic ties of utilitarian interests and uniform law (Gusfield 1975, pp. 4–5).

Where writers differed was in their reaction to the movement from traditional to modern, from community to society. For some the "loss" of community was a lamentable feature of the period. For traditionalists most especially Comte perhaps, the new order of economic logic, political interest groups, mass electorates and exchange-based market relationships represented the destruction of a stable social environment and traditional social hierarchies essential in their view to human organization. Others, however, viewed the change in a more positive light. Writers as different in other respects as Spencer and Marx saw community as coercive, limiting or idiotic.[3] In its place they saw the possibility of human equality and economic affluence as society overcame the limits and constraints intrinsic to community.

But whether nostalgic or ecstatic these theorists shared a belief in the "eclipse" of community in 19th-century Europe and America. Some of them, however, were careful in how far they would take the point. Weber (1949), for example, was quite specific in arguing that ideal-types such as community or society are not averages or descriptions, but models, useful in understanding reality but not to be mistaken for it. In current usage the concepts of community and society are opposites in an almost zero-sum form. This has involved reifying what started out as ideal-types or analytical terms into empirical descriptions (Bender 1978, pp. 32–40). There are several areas of contemporary social science in which this usage of the community–society distinction is most apparent. One is the study of "development" in "newly emerging" but economically "underdeveloped" countries. The other is political sociology.

The area of development studies, which includes work in various disciplines, has made considerable use of the concepts of tradition and modernity in attempts to understand and assess contemporary change. One particularly influential writer, Redfield (1930, 1955), described a contrast between ideal-type "folk" communities and "urban" societies. Although Redfield was careful to note the empirical coexistence of the two forms, his and similar work was taken up by a later generation of writers motivated by an urge to make their efforts relevant to the post-World War II world of new nations and the decline of European colonialism and less careful about using ideal-types (Lerner 1958, Almond & Coleman 1960, Rostow 1960). The problem was how could new nations overcome poverty and political instability to achieve economic growth and political stability? Equipped with the evolutionary perspective permeating modern social science they wrote of modernization, educational development, and political development. Each of these processes involves as a central moment

14

the total movement away from community to society. The pitting of one against the other, however, overlooks the mixtures or blends characteristic of actual settings. Moreover, it becomes an ideology of "antitraditionalism, denying the necessary and usable ways in which the past serves as support, especially in the sphere of values and political legitimation, to the present and the future" (Gusfield 1967, p. 362).

The "nation-building" focus of the development theorists is also an important element in orthodox political sociology. Figures such as Lipset and Almond are indeed important contributors to both literatures. Lipset's (1960) widely quoted essay, "Economic Development and Democracy", is a classic in the genre of modernization theory as well as an important background chapter of *Political Man*, a central statement of orthodox political sociology. Elsewhere and more recently Lipset (1981) tries to explain "backlash politics" or the rebirth of protest politics in Europe and the United States as a "revolt against modernity".

Whereas many 19th-century theorists bemoaned the passage of the past in the form of community, the modernization theorists find in this a welcome transformation. It is now prescribed as the means to overcome backwardness, deprivation, and penury. The evolutionary sequence suggests that the communal social system *must* retreat and give way before the impetus of the modern if modernity is to be achieved. In its present use, therefore, "community" is again an ideological counterpoint to actually existing societies. To writers such as Comte or Tönnies, community signified a world that was lost, whose disappearance highlighted the tragedy of modernity. In modernization theories it is a counter once more; but now as a world, if not yet eclipsed, to be eradicated in the interest of "modern society".

A number of writers have pointed up the centrality of discourse based around the community–society dichotomy in modern social science (e.g., Tilly 1973, Gusfield 1975). But they have also suggested its essential limitations. Tilly (1973), for example, notes that communities, by which he means "any durable local population most of whose members belong to households based in the locality" (Tilly 1973, p. 212), have not declined absolutely. Rather, as *places*, these have persisted in importance. But there has been a *relative* decline of such communities as bases of collective action: "local ties have diminished little or not at all, extralocal ties have increased" (Tilly 1973, p. 236). Society is experienced in place and community remains a *part*, if a diminished part, of this. Tilly concludes:

> One can notice the rising population of extralocal contacts and regret it. One can compare present conditions of solidarity with an ideal integrated folk community and find the present wanting. But on the basis of present evidence one cannot claim that urbanization [or modernization] produced an absolute decline in community solidarity.

Of particular importance and beginning in the 16th century, the formation of national states, the growth of international economic markets and industrial concentration have clearly led to massive social change (Tilly

15

1979). In consequence places have undergone fundamental change in the *relative* importance of local as compared to extra-local ties.

But to a considerable degree much empirical work in social science rests upon an a priori belief in the eclipse of community and the declining significance of place. Having confused place and community, place has then been defined as characteristic of communal association. The supposed eclipse of community has in turn led to the eclipse of place. Thus has orthodox social science effectively and systematically devalued place as a concept relevant to our time. Its association in the academic mind with parochialism and localism has become so deep-rooted that the idea of place as the structuring or mediating context for social relations seems strange and out of temper with the national-society focus of most contemporary social science.

Explaining the devaluation of place in orthodox social science

Why has the intellectual devaluation of place occurred? Hints at an explanation have appeared earlier. But it is an important and useful task to systematically sort out a "chain of explanation" from intellectual to social–historical sources of devaluation in orthodox social science.

The argument with respect to orthodox social science proceeds as follows: first, the most immediate intellectual source of devaluation is seen as the focus on the evolutionary sequence of transition from community to society. Second, an important correlate of the evolutionary perspective is identified. This is that the transition is natural, lawful, and universal. Third, the key historical period for the intellectual devaluation of place is located in the 19th century, an important period in the growth of nationalism as a place-transcending ideology. Fourth, it is argued that the extension of placeless social science in the form of modernization theory occurred in the historical context of the Cold War and the "struggle for hearts and minds" between "the West" and "World Communism" in the Third World of developing nations.

As previously argued, at the root of the intellectual devaluation of place in orthodox social science is the oppositional dichotomy of community and society and its image of total temporal discontinuity between two totally different forms of human association. In this usage, the concepts of community and society embody a specific theory of social change, that of the evolutionary transformation of life from community *to* society. When community and society are seen as opposites, co-existence is transitory and life must move from one end of the continuum to the other. In this process as society displaces community, place loses its significance since it is closely intertwined with community.

In bare essentials, therefore, the evolutionary sequence of movement from community to society provides the intellectual backbone to the devaluation of place in orthodox social science. But though a necessary element in a complete examination this is not in itself sufficient. In particular, there is no reference to why this approach should have become so compelling and why it did so when it did.

An important feature of late 19th-century social thought was its use of ideas from the natural sciences of the day. Evolutionary ideas from biology, particularly those of Lamarck and Darwin, were especially attractive to social thinkers. Motivations for drawing on these fields were mixed but one powerful encouragement was that natural explanations, ones drawing by analogy or homology from the natural sciences, would be free of the tinge of religion, ideology, and "free opinion" that had previously characterized social thought. Such explanations would also be universal and thus congruent with the tenets of the empiricist philosophy predominant amongst "scientists" and philosophers at that time.

Use of the term "natural" is potentially confusing in the context of a discussion of community and society. Proponents of society generally maintain the usage of the Physiocrats and have in mind unconstrained and self-regulated social order such as that found in nature. Proponents of community, usually critics of the status quo and thus easily portrayed as an ideological lot, use natural more in the sense of simple, primitive, or close *to* nature. In both cases, however, the former "scientific" and the latter "ideological," nature is invoked as the appropriate standard and grounding for explanation.

One can trace the scientific analogy to nature amongst social thinkers back at least to the 17th century. Many Enlightenment thinkers, perhaps especially the Physiocrats, aspired to explanation of social facts in terms of natural processes or by analogy to natural processes (Weulersse 1919, Fox-Genovese 1976). Viner (1960, p. 59) has noted of the Physiocrats that they

> arrived at their laissez-faire doctrine by way of a curious blend of the myth of a beneficent physical order of nature, of Hobbesism, of Cumberland's and Cartesian rationalism, and of some fresh and important economic analysis of the coordinating, harmonizing, and organizing function of free competition. There was a providential harmonious and self-operating *physical* order of nature, which, under appropriate social organization and sound intellectual perception, could be matched in its providential character, in its automatism, and its beneficence, in the *social* order of nature.

The mimesis or imitation of scientific discourse in the context of social theory encouraged a strong tendency towards objectification. Instead of sacred symbols and rituals as the expression of collective existence the essence of society is seen to be production and subsistence. Society is reconstituted in terms of functional relationships between *individual* people and things. Indeed people become indistinguishable from things. At the same time conceptions of natural law exclude the possibility of exception or idiosyncrasy. Natural law is associated with recurrent regularities, uniformities among phenomena, and classifications which omit singularities or differences (Gierke 1934, 1958, Crocker 1959, Meek 1976, Wade 1977).[4]

In this intellectual milieu social science found its first distinctive voice not as a new theory of society or politics but as a discourse about economics. Although economic ideas have roots that can be traced back to Aristotle's discussion of *Oikos*, one source of modern economics lies with the Physiocrats. Their ideas were later absorbed into a number of different theories. Physiocracy, *physis* (nature) + *Kratos* (rule), can be translated as nature's regime or as that form which embodies nature's rule. The Physiocrats discovered the economy to be a submerged order, suffocating under the burden of government and mercantilism, diverted from its natural course. The economy was thus an immanent order, ready to emerge and function according to laws inherent in the nature of things but stifled by politics (Meek 1976).

In asserting that free economic activity was "natural", the Physiocrats transferred to the plane of economics the cosmological understanding of a universe in which phenomena, social as well as physical, obeyed the laws of their nature, a nature perhaps initially implanted by God but now operating as a natural necessity without need of an "external" power to cause or correct it (Meek 1976, Livingstone 1984, Hamowy 1987). The freeing of the economy, therefore, became equivalent to the liberation of nature.

This is not to say that 19th-century theorists were of the same mind as the Physiocrats or one another when it came to either their views of *laissez faire* or nature or how to cast their analogies (compare Spencer and Marx, for example) (Anchor 1967). But it is to argue for a fundamental continuity in the attractiveness to many social theorists of the analogy to nature as essential to *scientific* explanation. Marx, for example, counterposed his scientific socialism to the socialism of those he characterized as utopians (e.g., Proudhon) largely on the presumed grounding of his argument in the natural order of things.

It is in the intellectual context of the naturalness and lawfulness of social life that the evolutionary perspective on the transition from community to society must be seen. Not only because the movement itself can be seen as natural and lawful but more especially because, at least in the hands of some theorists, the world of society is also natural and lawful when compared to the world of community. This implies, of course, that the analogy to nature is *always* social or political rather than scientific. In Gusfield's (1975) terminology concepts such as society or community are "utopias" as well as ideal-types. The active pursuit of society, however that concept is rendered, is justified as one with nature. How can nature be doubted?

In the 19th century the analogy to nature favored by many social theorists had perforce to co-exist with the growth of nationalism as a powerful political ideology. This led to an interesting convergence between the two (Mandrou 1973). Society, rather than remaining an abstraction or ideal-type, became coterminous with the boundaries of national states . . . and naturally so. Herder (in Barnard 1969) provided a clear connection and many social theorists followed suit: "a nation is as natural a plant as a family, only with more branches". A principle of what Smith (1979, p. 191) calls "methodological nationalism" came to prevail.

The three disciplines that were the major winners in the institutionalization of the study of social life, sociology, political science, and economics, each developed a subservience to the state. Indeed each had at its origins the practical interests of the state respectively, in social control, state management, and the national accumulation of wealth (Robinson 1962, pp. 24–25, Letwin 1963, p. 217, Nisbet 1966, p. 17, Deans 1978, p. 203, Skinner 1978, p. 350). At their roots, therefore, they were national in focus.

Many of the most influential thinkers of the 19th century were methodological nationalists. Whatever their other differences, Marx, Durkheim, and Weber all accepted state boundaries as co-extensive with those of the "societies" or "economies" they were interested in studying. More particularly, to some thinkers such as Durkheim the modern state was both creator and guarantor of the individual's *natural* rights against the claims of local, domestic, ecclesiastical, occupational, and other secondary groups. The state could thus function both as an "enforcer" of natural order as well as a "container", through the territorial definition it provided and the "statistics" it collected, of empirical observations about social and economic processes.[5] The categories used by the state for collecting statistics came to be the main operational categories of empiricist social science. Social science became "operationally" confined to national categories even when it looked elsewhere – to local settings or abroad.

Other disciplines such as ethnography, anthropology, and human geography served the state through the knowledge they provided about potential and actual "colonial societies" (Ranger 1976). These were ones in which community (and place) were still important. At "home", however, these disciplines, especially human geography, provided a rationale for territorial delimitation of the state (so-called "natural boundaries") and, notoriously, in the form of *geopolitik*, a rationale for state expansionism (see Kasperson & Minghi 1969, Parker 1982).

This is not to say that there were not real changes in the power of the state *vis-à-vis* other institutions. Great social and political change did occur in the 19th century. Above all the 19th-century state brought about the conditions for a liberal capitalist economy. In one way or another the development of national labor and commodity markets, the shift to factory production, the class segregation of urban populations and numerous changes in the texture of social life followed from this (Desai 1986). In the intellectual arena the growth of liberalism undermined the political and legal legitimacy of all groups and institutions that held an "intermediate" position between what we now think of as the "sphere" of the individual and that of the state (Frug 1980, p. 1088).

It was in this context that the social sciences became oriented towards the national state as a "natural" unit upon which to build their claims to generalization. This presupposed the diminished importance of community on a local scale and the social significance of place along with it. Even attempts at "reviving" community, perhaps most visibly in the hands of national dictators such as Mussolini and Hitler, but also more generally, have tended to operate solely at the national scale (Mosse 1975, Hobsbawn & Ranger 1983). Reference to local settings or to global processes was largely closed off by the nationalizing of social science and its subservience to the national state.

In his study of early 19th century America, Tocqueville (1945, II, pp. 14–18) argued that "democratic" nations have a greater capacity for general ideas than "aristocratic" ones. In abandoning the categories of the *ancien régime*, people were left to invent new categories with which to explain and give meaning to their experience. As he put it,

> When I repudiate the traditions of rank, professions and birth, when I escape from the authority of example to seek out, by the single effort of my reason, the path to be followed, I am inclined to derive the motives of my opinions from *human nature* itself, and this leads me necessarily, and almost unconsciously to adopt a great number of very *general* notions [my emphasis].

Pletsch (1981) has suggested that Tocqueville's reference to the propensity for general ideas provides an important perspective on the proclivity of social scientists to debilitate public discourse by imposing simple typologies on a complex social reality. One simple typology he selects for investigation is the taxonomic idea of three worlds that many of us use to divide up the world and its inhabitants. This "instance of primitive classification" is used by Pletsch to illustrate an intersection between politics and social science in the 1950s and 1960s that gave a boost to modernization theory and thus, from the present viewpoint, contributed to the intellectual devaluation of place.

In the aftermath of the World War II, Pletsch (1981, p. 569) argues:

> a great variety of social scientists and even journalists from several different nations, diverse ideological perspectives, and academic disciplines suddenly found the idea of a third world useful for organizing their thinking about the international order that had emerged from the settlements and unsettlements of World War II. And understanding the new order, they could place the significance of their own particular research in the grand enterprise of understanding social phenomena in general.

He traces the origins of the term Third World to a pair of binary distinctions. First, the world is divided into "traditional" and "modern" parts. Second, the modern part is divided into "communist" and "free" portions. These terms derive their meaning from their mutual opposition rather than from any relationship to real-world phenomena. The traditional world is the Third World, for example, whatever the situation "on the ground". Thus,

> The Third World is the world of tradition, culture, religion, irrationality, underdevelopment, over-population, political chaos, and so on. The second world is modern, technologically sophisticated, rational to a degree, but authoritarian (for totalitarian) and repressive, and ultimately inefficient and impoverished by contamination with an ideologically motivated socialist elite. The first world is purely modern, a haven of science and utilitarian decision making, technological, efficient, democratic, free – in short, a natural society unfettered by religion or ideology (Pletsch 1981, p. 574).

The distinctions are not only ideal-types, they also imply a historical relationship between the three categories they generate. Traditional societies are all destined to be modern ones somehow and to some degree. That has been the trajectory of history. That is how the First World has emerged. The Second World is an unfortunate perversion of "natural" modernization. The Third World is presently the zone of competition between these two models of modernity.

To Pletsch, the Cold War context of the 1950s and 1960s is the vital backdrop to the emergence of the social science schema of three worlds. Not only because of the boost it gave to modernization theory and the conventional teleology of "from community to society". More especially that even

> our strange taxonomy of social sciences – the belief there exist such distinguishable things as politics, economics, and society, and that there should be a separate group of social scientists to study each – has been subordinated in the last three decades to the idea of the three worlds (Pletsch 1981, p. 567).

As a consequence a new academic division of labor emerged. Thus,

> one clan of social scientists is set apart to study the pristine societies of the third world – anthropologists. Other clans – economists, sociologists, and political scientists – study the third world only insofar as the process of modernization has begun. The true province of these latter social sciences is the modern world, especially the natural societies of the West. But again, subclans of each of these sciences of the modern world are specially outfitted to make forays into the ideological regions of the second world (Pletsch 1981, p. 579).

This has served to reinforce the view of Western societies as the ones most amenable to nomothetic social science and consequently the ones in which reference to place is particularly inappropriate.

Tocqueville, then, usually read only as philosopher of early 19th-century American localism, provides a useful insight into why "modern" American social scientists, through their proclivity for "general ideas", have been attracted to a view of the world in which place is unimportant if not invisible. In the natural, democratic and economically free nations of the West, "the Free World," the concept of place *can have no role whatsoever.* They are now "beyond place," so to speak. The nation and natural law rule it out.

The devaluation of place in antithetical social science: the "logic" of liberal: capitalism and place

So far the intellectual devaluation of place has been accounted for in terms of a particular set of intellectual predispositions and contributory historical–political factors. This certainly fits the bill for orthodox social science. But why is it that antithetical social science, in particular Marxian political economy, usually opposed to the major tenets of the orthodoxy, has arrived at a similar devaluation of place? Of course, this viewpoint shares the evolutionism and naturalism of the orthodoxy. But thereafter they part company.

The major focus of Marxian political economy is on the logic of liberal capitalism. This logic postulates that most activities and goods are of no or limited value if they cannot be used in exchange. By distinguishing exchange-value from use-value, the value in exchange from the value in use of an activity of good, it is possible to argue that "pure exchange value" through the medium of money provides the basis for the practical devaluation of place. This is so because the process of circulation of commodities only reaches an intensive level if exchange value is detached from goods and exists as a commodity in its own right, i.e., money. Money makes possible the circulation of exchange-values across large distances and, in turn "makes possible the disembedding of social relationships from communities of high presence-availability" (local ties) (Giddens 1981, p. 115).[6]

More especially, in *capitalist* society the imperative of capital accumulation stimulates both technological innovation and geographical expansion, and thus the "universalization" of capitalism. The "passionate chase after exchange-value" (Marx 1967, p. 153) takes on global dimensions and leads to a world market that is in its turn the harbinger of a "world society". In Balbus's (1982, p. 42) words: "Capitalism, unlike any mode of production before it, provides the material basis for that universal, worldwide human solidarity that is foreshadowed in Marx's concept of species-being." But first, of course, those who produce (workers), being less committed to capital accumulation than those who own (capitalists), will throw off the chains of dominance. As a *universal* class the working class will both liberate itself and humanity in general.

The treatment accorded place in this intellectual tradition reflects a focus on the practical process under capitalist relations of production of "freeing" people from places and rationalizing such activities in terms of economic efficiency and market rationality. At the "deepest" level, so to speak, the intellectual devaluation of place in Marxist political economy is a product of an analysis that *accepts*, indeed welcomes, the logic of liberal capitalism and its associated features of individualistic atomism and commodification of people and places. Devaluation is economic, not "merely" intellectual. For all of Marx's invective against the "bourgeois" economy, he welcomes the social world that this economy has produced (Balbus 1982).

Giddens (1981, pp. 109–28) provides a useful analysis of what he calls the "money–commodity–money" relation and its implications for place. He draws on Marx's *Grundrisse* and Simmel's *The Philosophy of Money*, however, to argue that all commodities have a *double existence*, as "natural product" and as exchange-value. First of all, there *is* a practical basis to the devaluation of place. This can be seen in the "detachment" of people from places in the form of the commodification of land and labor (Pred 1983). Giddens (1981, p. 118) argues:

> Labour-power, as a commodified form, relates to labour in the traditional sense much as money relates to the use-values of goods for which it is exchanged. As a commodity, labour has a similar double existence to other commodities, as on the one hand the expenditure of human skills and abilities, and on the other a "cost" to capital, defined in terms of its value in exchange.

An important consequence of this is the conceptualization of people as a "factor of production," labor, on a par in the economic process to land and capital. People and places then become commodities; people as labor and places as locations.

However, and this is the main point in the context of this chapter, the "double existence" of labor-power as a commodity provides an important barrier to the "inexorable" logic of capitalism as the motor of modern history in two ways (Giddens 1981, p. 120). Firstly, labor-power is the only commodity which itself produces value. Secondly, the "other" aspect of labor-power (its non-commodity form) "is not merely the use value of a material good but a living human being with needs, feelings and aspirations. Labour-power is a commodity like any other – but resists being treated as a commodity like any other." Consequently, "labour-power consists of the *concrete activities* of human beings, working in *definite industrial settings*, who resist being treated as on a par with the material commodities which they produce" (my emphasis).

Place has been transformed, therefore. It is not just a question of simple solidarity by local reference groups in the face of "relations of absence," or "outside" influence, although this continues. Rather, to quote Giddens (1981, p. 121) again,

> The vast extension of time–space mediations made structurally possible by the prevalence of money capital, by the commodification of labour and by the transformability of one into the other, undercuts the segregated and autonomous character of the local community of producers. Unlike the situation in most contexts in class-divided [but non-capitalist] societies, in capitalism class struggle is built into the very constitution of work and the labour setting. In the context of the productive organisation, whatever sway the wage-worker gains over the circumstances of labour is achieved primarily through attempts at "defensive control" of the work-place.

Class relations are thus structured through the *place-based* exigencies of commodification *and* resistance.

In addition, however, the system of commodification institutes scarcity. When production and distribution are arranged through the behavior of prices and livelihoods depend upon getting and spending, insufficiency or shortage becomes the major datum for all economic activity (Sahlins 1972,

p. 4). In the presence of material abundance scarcity becomes a function of setting boundaries. As Hyde (1983, p. 23) argues:

> If there is plenty of air in the world but something blocks its passage to the lungs, the lungs do well to complain of scarcity. The assumptions of market exchange may not necessarily lead to boundaries, but they do in practice. When trade is "clean" and leaves people unconnected, when the merchant is free to sell when and where he will, when the market moves mostly for profit and the dominant myth is not "to possess is to give" but "the fittest survive," then wealth will lose its motion and gather in isolated pools. Under the assumptions of exchange trade, property is plagued by entropy and wealth can become scarce even as it increases.

Moreover, as Massey (1984) has argued in considerable empirical detail, "geographical variation" in economic activities "is profound and persistent" (p. 117). In particular, "Spatially-differentiated patterns of production are one of the bases of geographical variation in social structure and class relations" (p. 117). But the "uniqueness of places" is not simply an "outcome". Many industries today locate different parts of their production processes in different places to take advantage of the historically sedimented and peculiar attributes of different places. One of these might be rates of worker unionization, another could be the availability of skilled workers. Rather than annihilating space, therefore, the "logic of liberal capitalism" has led to a permanent but dynamic process of uneven development.

Commodification of land and housing reinforces this process. Rather than obliterating place, commodification of housing has stimulated an intense concern by property owners in particular over local events that threaten established use- and exchange-values (Agnew 1981). More generally, all people are trapped in a struggle over the appropriation of all values, as geographical patterns of external costs and benefits change in response to shifts in investment. Cox (1985) refers to this as the "politicization of location" from which has emerged a "spatial politics" based upon competition between places for new investment and a positive balance of external costs and benefits (also see Cox 1978, Logan 1978).

Thus, rather than disappearing in the face of the logic of liberal capitalism the *nature* of place has changed. The *balance* between "relations of presence" (local ties) and "relations of absence" (extra-local ties) has moved in the direction of the latter. But this has not led to the demise of place. Capitalism, while transforming society, has created a *new* structuring role for place (cf. Tilly 1973, 1979).

In *practice*, therefore, the logic of liberal capitalism has been both limited and resisted and has created new pressures for place. In "classic" Marxian political economy this has not been recognized. Rather, as Appleby (1978, p. 21) has pointed out in her discussion of the origins of economic thought:

> Abstractions describing commercial transactions had become more real in men's discourse than the tactile and concrete context in which they happened.

24

The lack of attention to the "double existence" of commodities has been particularly misleading. It has led directly to the view characteristic of orthodox economists, but shared by most theorists of "modern", "industrial", or "capitalist" societies, that concepts such as place are vestigial, useful perhaps for examining "past" societies but irrelevant to present or future ones. The focus on commodification has also missed the point that *dominant* social orders do not preclude the survival or creation of alternate ones in which other processes mediated in place are operative (Williams 1980).

Conclusion

I have traced the intellectual devaluation of place characteristic of orthodox social science to a combination of influences: sociological, political and intellectual. A static view of place, in particular its close association with community, has led over the previous century to its intellectual stillbirth. This reflects in part the enshrinement of the community–society metaphor as a major model of social change in orthodox social science, its "naturalization" as a scientific explanation and "nationalization" as a political explanation. The growth of modernization theory as a weapon in the ideological combat of the Cold War added another nail to place's intellectual coffin.

A powerful impetus to the devaluation of place has also come from "antithetical" social science in the form of Marxian political economy. In this literature there has been a tendency to absolutize the power of commodification and thus to see little scope for place in contemporary society or by extension in contemporary social science. The arguments of Giddens (1981) and others suggest, however, that though changed and changing in terms of the social relations it structures, place has a continuing relevance for understanding the workings of modern, capitalist society. Both orthodox and antithetical social science, therefore, have disposed of place peremptorily. It is past time for its rehabilitation.

Notes

1 The image of community as a harmonious unity was a reaction to the disruptions of the industrial era. Thinkers looked backwards without any sense that hierarchical power, class struggles, and civil wars were as characteristic of medieval or earlier times as of the period in which they lived (Braudel 1967).

2 Ancient writers, especially Aristotle, were drawn upon to lend support to the idea of movement from *Gemeinschaft* to *Gesellschaft*. However, as Springborg (1986) argues, Aristotle had a contractarian view of relations in the polis. To Aristotle *Gesellschaft* already existed in the urban Mediterranean society of antiquity.

3 Some remarks in the *Grundrisse* and *Capital* suggest that Marx could have formed a view of ancient society close to Aristotle's – one without the *Gemeinschaft/Gesellschaft* dichotomy. However, Marx seems to have fallen under

25

the influence of scholars such as Morgan, Maine, and Fustel de Coulanges so that in his later work he became an adherent of the *Gemeinschaft/Gesellschaft* myth (Springborg 1986, p. 202).

4   One important if neglected contribution of Enlightenment thinkers was the development of a skeptical scientific imagination (Szymanski & Agnew 1981, Seidman 1983). Moreover, the Enlightenment is only one, if a major, source of modern social theory. But in its contractarian-individualistic conception of society and as a result of the romantic-holistic reaction against this (e.g., as in Hegelianism) the Enlightenment's heritage has been more doctrinaire than many of its originators may have intended. In particular, the critique of contract theory that Seidman (1983) finds in *some* Enlightenment thought has not been very influential in modern Anglo-American social science though there have been some European thinkers, for example Simmel and Gurvitch, for whom this has been important.

5   Cohen (1982) explicitly relates the problem of enforcing national order to the collection of statistics. As she puts it (1982, p. 45): "Censuses pinned people down to a time, place and class; life tables suggested a numerical idea about the stages of life one could expect to live through". Porter (1986, p. 27) sees the enthusiasm for statistics in the early 19th century as an attempt "to bring a measure of expertise to social questions, to replace the contradictory preconceptions of the interested parties by the certainty of careful empirical investigation . . . the confusion of politics could be replaced by an orderly reign of facts."

6   Another way of seeing this is to focus on the "defamiliarization" of objects and symbols as a result of commodification. As J-C. Agnew (1983, p. 71) expresses it:

> When production becomes production for exchange, markets grow both more intensive and extensive. The commodity form moves outside its traditionally designated enclaves, with a corresponding impact on culture as a whole. Spheres of exchange collapse into one another; the gap between purchase and sale widens; and the problem of identity and accountability intensifies as buyers, sellers, and their goods grow in anonymity. In the measure that society itself becomes a marketplace, social relations, as one historian puts it, lose their "transparency".

References

Agnew, J. A. 1981. Home ownership and identity in capitalist societies. In *Housing and identity: cross-cultural perspectives*, J. S. Duncan (ed.). London: Croom Helm.

Agnew, J. A. 1987. *Place and politics: the geographical mediation of state and society.* London: Allen & Unwin.

Agnew, J-C. 1983. The consuming vision of Henry James. In *The culture of consumption: critical essays in American history, 1880-1980*, R. W. Fox & J. J. Jackson Lears (eds). New York: Pantheon.

Almond, G. & J. S. Coleman (eds) 1960. *The politics of the developing areas.* Princeton: Princeton University Press.

Anchor, R. 1967. *The Enlightenment tradition.* New York: Harper & Row.

Appleby, J. 0. 1978. *Economic thought and ideology in 17th century England.* Princeton: Princeton University Press.

Balbus, I. 1982. *Marxism and domination.* Princeton: Princeton University Press.

Barnard, F. 1969. *Herder on social and political culture*. Cambridge: Cambridge University Press.

Bell, C. & H. Newby 1971. *Community studies: an introduction to the sociology of the local community*. London: Allen & Unwin.

Bender, T. 1978. *Community and social change in America*. New Brunswick: Rutgers University Press.

Braudel, F. 1967. *Capitalism and material life, 1400–1800*. New York: Harper & Row.

Cahnman, W. J. 1977. Toennies in America. *History and Theory* **16**, 147–67.

Calhoun, C. J. 1978. History, anthropology and the study of communities: some problems in Macfarlane's proposal. *Social History* **3**, (3), pp. 363–73.

Calhoun, C. J. 1980. Community: toward a variable conceptualization for comparative research. *Social History* **5**, (1), pp. 105–29.

Cohen, P. C. 1982. *A calculating people: the spread of numeracy in early America*. Chicago: University of Chicago Press.

Cox, K. R. 1978. Local interests and urban political processes in market societies. In *Urbanization and conflict in market societies*. K. R. Cox (ed.). Chicago: Maaroufa Press.

Cox, K. R. 1985. The urban development process and the rise of the new spatial politics. Unpublished paper, Ohio State University.

Crocker, L.G. 1959. *An age of crisis: man and world in eighteenth century French thought*. Baltimore: Johns Hopkins University Press.

Deans, P. 1978. *The evolution of economic ideas*. Cambridge: Cambridge University Press.

Desai, M. 1986. Men and things. *Economica* **53**, pp. 1–10.

Durkheim, E. 1904. Review of A. Allin *The basis of sociality*. *L'année sociologique* 7 pp. 185.

Durkheim, E. 1933. *De la division du travail social*. 2nd edn. Paris: Alcan.

Faris, R. E. L. 1967. *Chicago Sociology, 1920–1932*. San Francisco: Chandler.

Fischer, C. S. 1975. The study of urban community and personality. *Annual Review of Sociology* **1**, pp. 67–86.

Fox-Genovese, E. 1976. *The origins of physiocracy*. Ithaca, NY: Cornell University Press.

Frug, G. 1980. The city as a legal concept. *Harvard Law Review* **93**, pp. 1059–154.

Giddens, A. 1981. *A contemporary critique of historical materialism*. Berkeley: University of California Press.

Gierke, O. 1934. *Natural law and the theory of society, 1500 to 1800*. Cambridge: Cambridge University Press.

Gierke, O. 1958. *Political theories of the Middle Ages*. Cambridge: Cambridge University Press.

Gusfield, J. R. 1967. Tradition and modernity: misplaced polarities in the study of social change. *American Journal of Sociology* **72**, pp. 336–51.

Gusfield, J. R. 1975. *Community: a critical response*. New York: Harper & Row.

Hamowy, R. 1987. *The Scottish Enlightenment and the theory of spontaneous order*. Carbondale, IL: Southern Illinois University Press.

Hays, S.P. 1967. Political parties and the community-society continuum. In *The American party systems: stages of political development*, W. N. Chambers & W. D. Burnham (eds). New York: Oxford University Press.

Hobsbawn, E. & T. Ranger (eds) 1983. *The invention of tradition*. Cambridge: Cambridge University Press.

Hyde, L. 1983. *The gift: imagination and the erotic life of property*. New York: Vintage.

Kasperson, R. & J. V. Minghi (eds) 1969. *The structure of political geography*. Chicago: Aldine.

Lerner, D. 1958. *The passing of traditional society.* New York: Free Press.

Letwin, W. 1963. *The origins of scientific economics: English economic thought 1660–1776.* London: Methuen.

Lipset, S. M. 1960. Economic development and democracy. *In Political Man.* London: Heinemann.

Lipset, S. M. 1981. The revolt against modernity. *In Mobilization, center-periphery structures and nation-building,* by T. Törsvik (ed.). Oslo: Universitetsforlaget.

Livingstone, D. N. 1984. Natural theology and neo-Lamarckism: the changing context of 19th century geography in the United States and Britain. *Annals of the Association of American Geographers* **74**, pp. 9–28.

Logan, J. R. 1978. Growth, politics and the stratification of places. *American Journal of Sociology* **84**, pp. 404–16.

Mandrou, R. 1973. *Des humanistes aux hommes de science: xvi, xvii siecles.* Paris: Seuil.

Marx, K. 1967. *Capital.* Vol. 1. New York: International Publishers.

Massey, D. 1984. Spatial divisions of labor: social structures and the geography of production. London: Macmillan.

Meek, R. 1976. *Turgot on progress, sociology and economics.* Cambridge: Cambridge University Press.

Mosse, G. L. 1975. *The nationalization of the masses: political symbolism and mass movements in Germany from the Napoleonic wars through the Third Reich.* New York: Howard Fertig.

Nisbet, R. A. 1966. *The sociological tradition.* London: Heinemann.

Parker, W. 0. 1982. *Mackinder: geography as an aid to statecraft.* Oxford: Clarendon Press.

Parsons, T. 1951. *The social system.* New York: Free Press.

Pletsch, C. E. 1981. The three worlds, or the division of social scientific labor, circa 1950-1975. *Comparative Studies in Society and History* **23**, pp. 565–90.

Porter, T. M. 1986. *The rise of statistical thinking, 1820–1900.* Princeton: Princeton University Press.

Pred, A. 1983. Structuration and place: on the becoming of sense of place and structure of feeling. *Journal for the Theory of Social Behavior* **13**, pp. 45–68.

Ranger, T. 0. 1976. From humanism to the science of man: colonialism in Africa and the understanding of alien societies. *Transactions of the Royal Historical Society* **26**, pp. 115–41.

Redfield, R. 1930. *Tepoztlan: a Mexican village.* Chicago: University of Chicago Press.

Redfield, R. 1955. *The little community.* Chicago: University of Chicago Press.

Robinson, J. 1962. Economic philosophy. London: Penguin.

Rostow, W. W. 1960. The stages of economic growth: a non-communist manifesto. New York: Cambridge University Press.

Sahlins, M. 1972. *Stone Age economics.* Chicago: Aldine.

Seidman, S. 1983. The Enlightenment and modern social theory. *Journal of the History of Sociology* **5**, pp. 66–90.

Shils, E. 1970. The contemplation of society in America. In *Paths of American thought,* M. White & A. Schlesinger Jr (eds) Boston: Houghton Mifflin.

Skinner, Q. 1978. *The foundations of modern political thought.* Vol. 2. Cambridge: Cambridge University Press.

Smith, A. D. 1979. *Nationalism in the twentieth century.* Oxford: Martin Robertson.

Springborg, P. 1986. Politics, primordialism, and orientalism: Marx, Aristotle, and the myth of Gemeinschaft. *American Political Science Review* **80**, pp. 185–211.

Szymanski, R. & J. A. Agnew 1981. *Order and skepticism: human geography and the dialectic of science.* Washington, DC: Association of American Geographers.

Tilly, C. 1973. Do communities act? *Sociological Inquiry* 43, pp. 209–40.

Tilly, C. 1979. Did the cake of custom break? In *Consciousness and class experience in nineteenth-century Europe*, J.M. Merriman (ed.). New York: Holmes & Meier.

Tocqueville, A. de 1945. *Democracy in America*. Rev. edn New York: Knopf.

Tönnies, F. 1887. *Community and association*. London: Routledge & Kegan Paul.

Viner, J. 1960. The intellectual history of laissez-faire. *Journal of Law and Economics* **3**, pp. 45–69.

Wade, I. O. 1977. *The structure and form of the French Enlightenment*. Princeton: Princeton University Press.

Webber, M. 1964. Urban places and the non-place urban realm. In *Explorations in urban structure*, M. Webber et al. (eds) Philadelphia: University of Pennsylvania Press.

Weber, M. 1925. *The theory of social and economic organization*. New York: Free Press.

Weber, M. 1949. Class, status, and party. In *From Max Weber*, H. H. Gerth & C. W. Mills (eds). London: Routledge & Kegan Paul.

Weulersse, G. 1919. *Le mouvement physiocratique en France, de 1756 à 1770*. Paris: Gallimard.

Williams, R. 1980. *Problems in materialism and culture*. London: Verso.

Wirth, L. 1938. Urbanism as a way of life. *American Journal of Sociology* **44**, pp. 1–24.

Zimmerman, C. 1938. *The changing community*. New York: Harper.

3

Place, region, and modernity

J. NICHOLAS ENTRIKIN

The concepts of place and region have occupied an ambiguous position in the conceptual landscape of 20th-century social science. Through this century, the study of regions has moved toward the periphery of social science and beyond. In this chapter, I will consider the manner in which American geographers and other social scientists have sought to fit regional studies within the prevailing conceptions of scientific rationality in the 20th century. The most common view of such attempts has been that they have been unsuccessful. The modern model of scientific rationality has been drawn from the physical sciences, and regional studies have not conformed with this model. This fact is an important one for understanding the history of geography in this century.

Before stating my central themes, I will clarify several terms. The study of place and region have generally been referred to as "chorology". I will adopt the language of the chorologists, and will therefore use the terms "place" and "region" such that, except for the differences in geographical scale, their meanings are essentially equivalent. Chorologists have employed two distinct meanings of these terms in referring to the "generic" and the "specific". Generic places are places of a certain type, and specific places are those defined in terms of their individuality or uniqueness. The concept of specific place has been the most distinctive aspect of attempts to create a human science of regions and has been the most controversial part of such attempts (Sack 1974). For example, the idea of specific place has been the basis for descriptions of regional studies as an idiographic science. I will be concerned most with the concepts of "specific place" and "specific region", and will use the terms "place" and "region" as a shorthand to refer to these ideas.

My theme will concern the underlying bases for judgments of the significance of regional studies. I will suggest that the criteria of significance have changed in relation to beliefs that we hold concerning modernity and modern rationality. The language of modernity that I will emphasize is that of science, for, as has been frequently noted, modern Western cultures have tended to make scientific knowledge synonymous with rationality. Our conception of science has changed throughout the century, however. In recent years historians and philosophers of science have questioned many of the assumptions of a positivistic or logical empiricist conception of scientific knowledge, a conception that came to dominate the social

30

sciences during the middle part of this century. Part of the legacy of this questioning is a recognition of both the multidimensional character and the social and cultural context of scientific knowledge. A consequence of this multidimensionality is that shifts in disciplinary orientation may originate at one of a number of different levels of analysis, for example in debates concerning matters of empirical analysis, methodology, theory construction, epistemology or metaphysics (Alexander 1982). Such debates are often carried on quite independently from those found at other levels, despite the fact that they have consequences for one another. The influence of the cultural and social context may be seen at each of these levels.

These two aspects of a post-positivist perspective on the history of social science are helpful in addressing the question of the changing significance of regional studies in the 20th century. The question of significance, which has a dual sense of both importance and meaning, can be divided into several conceptually distinct yet related parts, empirical-theoretical significance, normative significance, and scientific significance (Weber 1949, pp. 131–63, Keat 1981, pp. 52–8). The first addresses the question of the areal variation of economy, society, and culture in modern societies; the second addresses the cultural value associated with this areal variation; and the third will more specifically address the question of the scientific nature of chorology. All such judgments eventually relate to a culture's changing conception of itself, a conception that in American thought can be linked in part to ideas on the nature of modernity. Modernity is an abstraction that, as sociologist Peter Berger describes, is rooted in underlying institutional processes such as a capitalist market economy, a bureaucratized state, a technologically advanced economy, and a mass communications media, but which is also associated with certain forms of consciousness, one of which is the idea of technical rationality as the sole form of rationality (Berger 1977).

Empirical-theoretical significance

One of the frequently noted characteristics of modern life and an assumption of most theories of modern society is that there has been a decline in the areal variation of ways of life during the 20th century. Many reasons have been offered to explain this decline, the most prominent of which are the technological revolutions in communication and transportation, the growth of a worldwide capitalist market system, the growth of urbanization associated with the structural transformation of economies from an agricultural form to an industrial and post-industrial form, and the growth of the state that is associated with the centralization of decision-making and the application of principles of scientific rationality to the planning of societies. These reasons are of course interrelated and have been combined to present a portrait of modern society. Dispute arises primarily about the relative weighting and the relationships among these factors, and not about their descriptive accuracy.

The first of these reasons, the 20th-century revolution in technology that allowed for increased ease and speed of movement of people and information, is most often found in discussions of chorology. For example, in the 1948 symposium on regionalism in America, the sociologist Louis Wirth presented a critical view of the significance of regions in modern life when he stated that:

> We must always, especially in modern times, reckon with the power of communication and transportation – with the mobility of men and ideas – to undo regions. (Wirth 1965, pp. 388–9)

This same idea was presented in a somewhat different fashion by the geographer Cole Harris who noted that the technological revolution meant that "isolation, the essential support of regional cultural differentiation, was breaking down" (Harris 1978, p. 123).

This breakdown of cultural "isolation" has been placed in a long-term perspective by those considering the evolution of a capitalist world system. These ideas have been widespread among historians and social scientists and thus cannot be assigned to the work of any one person, but a significant contributor to contemporary discussion of this issue has been the sociologist Immanuel Wallerstein in his consideration of the process of modernization (Wallerstein 1974, 1979). Modernization is associated with the origins of the world-system of the capitalist mode of production that he locates in Europe during the period from 1450 to 1640. After a period of consolidation that lasted until the 19th century, this mode of production expanded into a global system through the aid of the powerful technologies unleashed by the industrial revolution. Wallerstein suggests that this period of globalization lasted until World War I, and from that time to the present has gone through a period of consolidation. The greater interconnectedness of this world system is powered by the market, and facilitated by the technological innovations that such a market economy has produced.

David Harvey has captured the geographical significance of this spread of capitalism in stating that:

> Peoples possessed of the utmost diversity of historical experience, living in an incredible variety of physical circumstances, have been welded, sometimes gently and cajolingly but more often through the exercise of ruthless brute force, into a complex unity under the international division of labour. (Harvey 1982, p. 373)

Harvey is not arguing here that the development of capitalism leads necessarily to the destruction of areal diversity. On the contrary, he would argue that the logic of capitalism provides a basis for understanding the process of differentiation through what has been referred to as "uneven development":

> Factories and fields, schools, churches, shopping centres and parks, roads and railways litter a landscape that has been indelibly and

irreversibly carved out according to the dictates of capitalism. Again, this physical transformation has not progressed evenly. Vast concentrations of productive power here contrast with relatively empty regions there. Tight concentrations of activity in one place contrast with sprawling far-flung development in another. All of this adds up to what we call the "uneven geographical development" of capitalism. (Harvey 1982, p. 373)

As his words suggest, however, and as Neil Smith has illustrated, the study of uneven development shares little with traditional studies of chorology other than a concern for areal variation. Harvey's language of "empty" and "filled" regions suggests a conceptualization closer to the language of the spatial analysts than to the chorologists (Smith 1984). The picture that emerges from his writing is the manipulation or creation of spaces and regions by the powerful mechanisms associated with the growth of capitalism. Ideas of local culture emerge only as residual effects of this dynamic of capital, and are to be understood only in reference to them.

The globalization of capitalism has been associated with industrialization. The settlement pattern that emerged with the concentration of industry after its initial period of dispersion was that of urbanization. The process of urbanization has been seen as destructive to traditional patterns of life and thus as destructive to the diversity of ways of life or cultural forms. It has also been associated with the standardization of landscapes ranging from the pattern of suburbanization of low density residential communities to the architecture of the so- called "International School" (Relph 1981).

The increased scale of production and the increased scale of social organizations as illustrated in the burgeoning metropolis created the need for ever more complex and centralized forms of organization and administration. It became clear from the late 19th century that the "invisible hand" of commercial regulation could not alone create the kind of social harmony and control necessary for the continued expansion of capital. Thus, as the political theorist Theodore Lowi suggests, "modern industrialized society can be explained as an effort to make the 'invisible hand' as visible as possible", through the application of principles of rationality (Lowi 1979, p. 15). Lowi notes that "rationality was used on markets as well as in them", and "came to be set above all other values" (Lowi 1979, p. 15).

A manifestation of this instrumentalist rationality has been the growth of the governmental bureaucracy and its expansion into most areas of modern life. This growth has been described as a contributory factor in the decline in regional differentiation in at least two ways. The first of these involves the centralization of decision making. For example, a stimulus to the regionalist movements in the 1930s was the reaction against the expanded role of the federal government with the programs of the New Deal. The second is the growth of local government. One would expect this growth to counterbalance the trend toward centralization, encourage local autonomy, and hence stimulate areal diversity. The prevalence of scientific principles of management and administration throughout this level of government has had a quite opposite effect, however. This fact is especially evident in the growth of local and regional planning in the 20th century. What began as an organization that *supported* local community and encouraged regionalism was transformed into an important force in the "rationalization" of landscapes (Friedman & Weaver 1979).

Descriptions of the processes of modernization and rationalization became the basis for policy and planning. For example, a concern with the rational organization of urban landscapes became part of the training of the professional planners whose numbers and powers have expanded significantly throughout the 20th century. With the authority warranted by society and given to its technical and scientific experts, the planner is able to manage and manipulate the environment to better fit with accepted standards of a rational landscape.

While we associate this rise of the professional expert with the rationalization process in modern Western society, this belief should be tempered by the comments of the philosopher Alfred N. Whitehead who noted that:

> Professionals are not new to the world. But in the past, professionals have formed unprogressive castes. The point is that professionalism has now been mated with progress. (Whitehead 1957, pp. 294–5)

It is this authority based upon the pre-emption of the vision of the progressive forces of modernity that Marshall Berman has offered as an explanation for the ability of legendary planner Robert Moses to destroy the neighborhoods of New York to make room for his massive "public works" projects. As Berman states: "Moses was destroying our world, yet he seemed to be working in the name of values that we ourselves embraced" (Berman 1982, p. 295). While Moses' works left a set of symbols of modernism impressive for their scale, the more mundane applications of modern planning techniques were notable for their tendency to contribute to the standardization of the urban landscape.

The goals of the original regional planning movement were indicative of the general concern with local autonomy and decentralization of decision-making that was an undercurrent of most 20th century American regionalist thought. These ideals were somewhat vague, and support for them could be found in widely variant political theories (Clark 1985, p. 196). They were, nonetheless, powerful ideals that had supported the study of regional diversity as well as the closely related communitarian movement. They were ideals that lent a normative significance to regional studies.

Normative significance

For some, the decline in the areal diversity of social life in the 20th century was a matter of great social and political import. For example, in American thought the regional diversity of ways of life was considered to be an important element in the maintenance of democratic institutions, and thus an aspect of life to be valued and maintained. This view is clearly evident in the writings of early 20th-century intellectuals such as the philosopher Josiah

Royce and the historian Frederick Jackson Turner. Royce's philosophy of provincialism was an attempt to bridge the divide between his highly abstract absolute idealist philosophy and the concerns of modern life that his colleague William James had so successfully captured in his pragmatist philosophy. Provincialism for Royce was the loyalty to place and to local community. Such loyalty was not simply an abstract principle, but was intended by Royce to function as a guide for action, one that would serve as a corrective to what Royce viewed as the leveling tendencies evident in modern society. Such tendencies included industrialization, urbanization, the increased mobility of the population, the development of mass media, and the growth of a mob spirit. The combined effect of these tendencies was the loss of provincial customs and ways of life, and the tyranny of the nation over the province (Royce 1908; see also Entrikin 1981).

The relation of province to nation was parallel to that of individual to community. Both the nation and the community derived strength and unity through diversity. The community was held together by the loyalty of independent individuals to a common purpose, rather than through a fusion of many minds into a single way of thinking. The first coincided with the republican ideals of civic community, and the second with Royce's view of the mob spirit. This same unity through diversity that was the ordering principle of the ideal community was also an essential quality for a strong nation state. Royce saw the ideal national order as one based upon a similar bonding of independent provinces with distinctive ways of life into a nation state.

The republican ideals of civic community that were evident in Royce's work were an important source for the normative significance of regional diversity. Another source was the concern for the understanding and the conservation of the diversity of life forms that has been a part of natural history and evolutionary biology. Ideas from these two distinct sources were drawn together in late 19th-century social theory, and were evident in the ecological and regional traditions of American social science. The best example of this combination was found in the work of Frederick Jackson Turner, especially in his discussion of sectionalism. Turner studied the natural sciences as well as history and sought to place history on a firm "scientific" basis. He did this through the consideration of geology, biology, and geography and their influences on civilization. As the historian Michael Steiner pointed out, however, Turner's sectionalist thesis "reached its final shape as Turner joined his environmental studies with Josiah Royce's concept of 'higher provincialism'" (Steiner 1979, p. 455). Steiner suggested that:

His [Turner's] most important discussion of sectionalism relies upon Royce's belief that a wise provincialism would nurture careful use of the earth and also support a sense of community amid mass society. (Steiner 1979, p. 456)

The closing of the frontier meant a gradual shift from the strong individualism that marked frontier life to a co-operative effort that builds community.

35

The wandering that marked frontier life would be replaced by the growth of settlement, the stability of a population and the development of local custom and identity. The preservation of this localism or sectionalism was an important component in the stability of the democracy that it supported. Turner referred to Royce in stating that:

> It was the opinion of this eminent philosopher that the world needs now more than ever before the vigorous development of a highly organized provincial life to serve as a check upon mob psychology on a national scale, and to furnish that variety which is essential to vital growth and originality. With this I agree. (Turner 1959, p. 45; cited in Steiner 1979, p. 463)

This concern for the beneficial aspects of attachment to place and local community as important elements in the stability of a democracy was a theme that was found also in the next generation of American scholars. I will illustrate briefly this fact by considering the theoretical perspectives of the sociologist and ecologist Robert Park, the regional sociologist Howard Odum, and the cultural geographer Carl Sauer. Each had quite distinct substantive interests, but they shared common naturalistic assumptions that served as both an implicit and explicit theory in their work. Each used the ecological principles of harmony, balance and equilibrium as the basis for their understanding of the ideal social order.

Park and Odum were the founders and the intellectual leaders of two prominent traditions in American sociology during the first half of the 20th century, Park of the ecological tradition at the University of Chicago and Odum of the regional sociology movement centered at the University of North Carolina at Chapel Hill. Their research interests converged on the issue of race relations, but diverged on many other topics, and had significantly different impacts on their field. Odum has been considered the more provincial of the two because of his overriding concern for the traditions and the progress of the South, and figures more prominently in studies of 20th-century Southern history and in studies of the history of regional planning than he does in histories of 20th-century sociological thought (Friedman & Weaver 1979, O'Brien 1979, Reed 1982, Singal 1982). His regional sociology essentially died with him (Reed 1982, pp. 33-44). The ecological tradition of Park survived beyond the lifetime of its founder, carried on first by Park's students and then later by geographers and sociologists (Berry & Kasarda 1977).

The differential impact of the ideas of the two men can also be seen in part as a reflection of their locations. Odum was a progressive in the South, a firm believer in the view that progress could be achieved by applying scientific principles to social life. The traditions and landscapes that he and his students studied were those of an older agricultural and rural social order. Park and his students, on the other hand, were describing the newly emerging modern order of the industrial metropolis. The ecologists sought general principles of the social and spatial order of this metropolis, but their general principles were more often abstractions derived from the industrial, pre-automobile city of early 20th-century America, or, more specifically, from Chicago.

These two sociologists saw a common focus in their concerns, related to their theoretical interest in the ecological aspects of the social order. In 1937, Park wrote to Harry E. Moore, the co-author with Odum of the book *American regionalism*, and noted that Moore was looking at the "matter of regionalism from the same point of view as myself" (Park 1937). Park expressed an interest in the "theoretical aspect of regionalism", part of which included the concern with "balance". Park's ecology posited a dual order of social life, the ecological and the moral (Park 1926). The first of these described a political economy of competitive individualism, an order that could be described in terms analogous to those used to describe the competition between plants in biotic communities. The second described the order of culture, a symbolically mediated order of shared meanings, values and goals. The legacy of this Chicago tradition has been most often associated with the study of the ecological order in that most of the studies of Park's students were based in these concerns. With the notable exception of Walter Firey's work, little was done in terms of empirical studies of the impact of the moral order (Firey 1968). And, even in his studies of land use in Boston, Firey tended to reduce the cultural to the level of the subjective and the epiphenomenal (Suttles 1984).

Park argued that the foundation of this dual order rested upon two intellectual sources. He described these in a letter to his colleague Roderick McKenzie as first, the study of the civic community and second, the study of anthropogeography (Park 1924; see Entrikin 1980). Odum's writings blended similar parts, but in different proportions.

Odum's writings can be placed chronologically among the second generation of American sociologists, but his ideas correspond more closely with those of the first generation (O'Brien 1979, p. 37). He saw society as an organic unit of diverse, yet interrelated, parts. The progressive development of an organism was directly related to harmony among its parts. Expressed in spatial terms, the progress of a national social unit was a function of the harmony of its regions. The concept of regional harmony or balance was a guiding idea in his work and one that he saw as having both theoretical and practical significance. In Odum's regionalist philosophy "...the South would become organically interconnected with the rest of the country to form a coherent 'integrated' whole while still retaining part of its identity" (Singal 1982, p. 149). Besides 19th-century organicist social theory, the primary sources of his thought were the regionalist ideals of Josiah Royce and Frederick Jackson Turner and the cultural anthropologists with whom he grouped the geographer Carl Sauer.

After a visit to Chapel Hill during the 1930s, Sauer wrote enthusiastically about the approach to the study of society taken by Odum and his colleagues. He compared them to a group of natural scientists as opposed to social scientists because of their keen sense of observation and their distaste for epistemology (Sauer 1938). Sauer viewed the North Carolina group as fellow culture historians. Culture history was defined by Sauer as the natural history of man. The guiding ideas and controlling metaphors of his studies were drawn from the natural sciences, more specifically those natural sciences such as biology and geology that have had a strong natural history tradition. Two of the most important of these guiding ideas concerned the issues of balance and diversity (Entrikin 1984).

For Sauer, the tendency toward the increased diversity of life forms was a part of the natural order. During the last several hundred years of human history, especially since the time of the industrial revolution, man has disturbed this natural order and has worked as an agent opposing this tendency toward diversity. These efforts have led to a world out of balance. The declining variety in the natural world was matched by a similar decline in cultural diversity. The growth of large scale economies and of concomitant bureaucratic states was at the expense of local community and the kind of cultural diversity that derives from autonomous local communities.

The works of these three social scientists, Park, Odum and Sauer, reinforce the tendency to view regional studies as an outgrowth of a naturalistic social theory so prevalent in 19th-century social thought. The common explanation of this phenomenon is the importance of Darwin and the ideas of evolutionary biology in late 19th-century thought, and also the attempts by social scientists to place the newly emerging social sciences on a firm scientific foundation by copying the methods and the vocabulary of the natural sciences. Recent work in the history of science suggests, however, that a more complicated relation existed between ideals of the natural order and those of the social order. In one such study of the history of ecology, J. Ronald Engel's *Sacred sands*, the author considers the interesting mix of social and political theory that infused the thought of early 20th-century animal and plant ecologists (Engel 1983; see also Tobey 1981). Such a mix indicates that the frequently stated assumption of the one way flow of ideas between the natural sciences and the social sciences requires re-evaluation.

The concerns of individualism, communitarianism, and the loss of social and cultural diversity in modern life are closely intertwined with the history of regional studies. A primary difference between social theory in the early 20th century and that of the late 20th century is the manner in which these themes have been considered. In the present day such themes are unlikely to be placed in relation to nature or to be joined through metaphor or analogy to the natural world. Through much of the 20th century, social scientists have abstracted these themes from place and space. The areal variation of society and culture thus no longer shares in the normative significance associated with these themes that have been central to the study of American society and culture.

Scientific significance

The scientific significance of the regional concept has been a troublesome issue in the geographical literature throughout the 20th century. Chorology did not conform to the standards of objective science, and its primary concepts, specific place and specific region, contradicted the rules of scientific concept formation. One of the sources of this difficulty has been the lack of theories or laws in the study of chorology. This view was expressed in 1937 by John Leighly, who argued that:

> There is no prospect of our finding a theory so penetrating that it will bring into rational order all or a large fraction of the heterogeneous elements of the landscape. There is no prospect of our finding such a theory, that is to say, unless it is of a mystical kind, and so outside the pale of science.
>
> There must be, to return to an earlier phase of the argument, selection among regions to be described as well as selection of items of information to be included in regional or topographic descriptions. But the regionalist position provides no logically given criteria for selection, save the most general one that the region selected exist on the earth. (Leighly 1937, p. 128)

For many regional geographers the tradition of a relatively standard format of presentation was often confused with a theoretical framework and thus served as the basis of judgment for selection.

This same criticism was raised two decades later and became the basis for the spatial analysts' condemnation of the chorological approach. Fred Schaefer, William Bunge, David Harvey, Peter Haggett and others criticized the regional approach as descriptive, subjective, and atheoretical and thus as unscientific (Schaefer 1953, Haggett 1965, Bunge 1966, Harvey 1969). The spatial analysts' commitment to the idea of a nomothetic science of geography became the foundation for the transition of reigning orthodoxies in human geography, from the study of the individual region to the search for general laws of spatial organization.

The logical centerpoint around which many of these arguments revolved was that of the neo-Kantian conception of chorology as an idiographic science. The legacy of those authors who have described geography as in part an idiographic science has been clouded by a number of enduring misinterpretations and ambiguities concerning studies of the particular. Included among these are the views that (a) idiographic studies were based upon the existence of unique objects and events, (b) idiographic studies did not use general concepts, and (c) idiographic studies did not seek causal explanations. If, however, one considers the original statement of the idiographic–nomothetic distinction by Wilhelm Windelband and its subsequent elaboration by Heinrich Rickert and Max Weber, it becomes clear that the terms refer to modes of concept formation that differentiate sciences and not to the content of the sciences. Also, these philosophers and social scientists recognized the importance of general concepts and causal relations in idiographic studies. These two issues were kept distinct, however, by the postulation of two types of causal relations, one between individual phenomena and the other between classes of phenomena (Entrikin 1985).

Geographers advocating a logical empiricist view of their field have generally placed the idiographic study of regions in one of two categories, either as unscientific description or as prescientific "explanation sketches".

Both these claims rest upon the recognition of the subjectivism inherent in chorology, a subjectivism that is attributed to the need to make selections and to judge significance on the basis of personal judgment rather than on the basis of tested generalizations, laws or theories. These two distinct views are derived from differing interpretations of the degree of subjectivity involved, and from differing levels of tolerance for subjective judgment in science. The more recent statements of the relation of chorology to a scientific geography have tended to be more conciliatory, and have regarded chorology as an early stage in the development of a nomothetic science. The ideas of Robert Sack best express this conciliatory tone. In Sack's view chorology can be placed along a continuum which has as its two endpoints the objectivity of the physical sciences and the subjective judgment of artistic creation. Chorological interpretation was placed in the middle of this continuum with historical narrative. Both were described as explanatory models that have a form very close to that employed in everyday discourse (Sack 1980, pp. 94-5).

The ambiguous relation of the idea of specific place with the rules of scientific concept formation has been illustrated in different terms by Sack's arguments concerning spatial concepts in differing cognitive modes. For example, he distinguishes the mythical conception of space from the scientific conception by suggesting that in the mythical conception experience is fused with geographic context. The experience and the place become one, and thus are conceptually inseparable. In the language of science experience and geographical context are separated, and place becomes simply the location of objects and events. The idea of specific place employed by chorologists has tended to be closer to the fused conception of myth than to the unfused conception of science (Sack 1980, p. 86). This conceptual fusion has contributed to our sense of the empirical and the normative significance of the ideas of place and region, but at the same time has made them seemingly unsuitable as scientific concepts.

Regional studies in the late 20th century

The failure of chorologists to justify their concepts in terms of prevailing standards of scientific concept formation provides a clue for understanding the recent interest in the concepts of place and region. In geography, this expression of interest has come primarily from those who have sought to redirect geographical research toward a concern for the richness of human experience and an understanding of human action. In carrying out their research projects they have called into question assumptions concerning life in modern societies as well as beliefs concerning the rationality of social science. This questioning is in part a function of the fact that they are taking seriously the cultural significance of everyday life.

Fifteen years ago, David Sopher noted that:

The greatly changed scale of interaction with the land, reach of information, and areal scope of culture that characterize complex, technologically sophisticated societies oblige the cultural geographer to change his own methods and scale of investigation in order to deal with them. (Sopher 1972, p. 334)

Only very recently, however, have cultural geographers begun to overcome the intellectual, institutional and professional "barriers" that have directed them to the study of traditional as opposed to modern societies. Their insights into modern societies have often concerned the role of traditional culture traits, and this emphasis has had both positive and negative effects.

On the positive side, cultural geographers have contributed to a growing body of literature in the social sciences that has recognized the persistence of traditional aspects of culture despite the processes of modernization. Theories of modernization have often neglected what Clifford Geertz has referred to as "primordial attachments" that "stem from the 'givens' – or, more precisely, as culture is inevitably involved in such matters, the assumed 'givens' – of social existence" (Geertz 1973, p. 259). Such attachments remain despite change. More specifically, attachment to place and to territory remain of importance in modern society despite the increased mobility of the population and despite the production of standardized landscapes. This significance is expressed in the language of everyday life through the conceptual fusion of geographical context and experience, and it is this language and its symbolic content that cultural geographers have drawn attention to after a long period of relative neglect in human geography. The negative implications derive from the tendency to characterize these elements of culture as part of the private world of individual subjectivity. Their manifestation and their study are seen as forms of escape from the "real world" of a rationalized and alienating social order, and from the explanations of this social order that are presented in terms of the technical rationality of social science. The elements of culture studied are judged as having little impact on the modern social system. A contributing factor in this assessment is that the study of these symbols, myths, and metaphors are not easily accommodated within our prevailing conceptions of the logic of social scientific inquiry.

However, the arguments of cultural geographers, cultural anthropologists, and cultural sociologists have raised questions concerning our understanding of modernity at the same time the post-positivist historians and philosophers of science have undermined the relatively narrow conception of rationality that has dominated discussions of the logic of social science. The result of this combination of intellectual activity has been a growing recognition of the problematic nature of the boundaries that have been drawn between subjectivity-objectivity, rationality-irrationality, and the demarcation of scientific knowledge from other forms of knowledge. In geography this recognition has been manifested in a renewed interest in questions related to what might be termed the existential core of geography, the fact that human experience is always rooted in place. For some geographers this represents a relatively recent interest. For David Sopher, it was at the center of a life's work.

41

References

Alexander, J. 1982. *Theoretical logic in sociology, Vol. 1, positivism, presuppositions, and current controversies*. Berkeley: University of California Press.
Berger, P. 1977. *Facing up to modernity: excursions in society, politics, and religion*. New York: Basic Books.
Berman, M. 1982. *All that is solid melts into air: the experience of modernity*. New York: Simon & Schuster.
Berry, B. & J. Kasarda 1977. *Contemporary urban ecology*. New York: Macmillan.
Bunge, W. 1966. *Theoretical geography*. Lund: C.W.K. Gleerup.
Clark, G. 1985. *Judges and the cities: interpreting local autonomy*. Chicago: University of Chicago Press.
Engel, J. R. 1983. *Sacred sands: the struggle for community in the Indiana dunes*. Middletown, CT: Wesleyan University Press.
Entrikin, J. N. 1980. Robert Park's human ecology and human geography. *Annals of the Association of American Geographers* **70**, pp. 43–58.
Entrikin, J. N. 1981. Royce's provincialism: a metaphysician's social geography. In *Geography, ideology and social concern*, D. R. Stoddart (ed.). Oxford: Basil Blackwell.
Entrikin, J. N. 1984. Carl O. Sauer, philosopher in spite of himself. *Geographical Review* **74**, pp. 387–408.
Entrikin, J. N. 1985. Humanism, naturalism, and geographical thought. *Geographical Analysis* **17**, pp. 243-7.
Firey, W. 1968. *Land use in central Boston*. Westport, CT: Greenwood Press.
Friedman, J. & C. Weaver 1979. *Territory and function: the evolution of regional planning*. Berkeley and Los Angeles: University of California Press.
Geertz, C. 1973. The integrative revolution: primordial sentiments and civil politics in the new states. In *The interpretation of cultures*. New York: Basic Books.
Haggett, P. 1965. *Locational analysis in human geography*. London: Edward Arnold.
Harris, R. C. 1978. The historical geography of North American regions. *American Behavioral Scientist* **22**, pp. 115–30.
Harvey, D. 1969. *Explanation in geography*. London: Edward Arnold.
Harvey, D. 1982. *The limits to capital*. Oxford: Basil Blackwell.
Keat, R. 1981. *The politics of social theory: Habermas, Freud and the critique of positivism*. Oxford: Basil Blackwell.
Leighly, J. 1937. Some comments on contemporary geographic method. *Annals of the Association of American Geographers* **27**, pp. 125–41.
Lowi, T. J. 1979. *The end of liberalism: the second republic of the United States*. New York: W.W. Norton.
O'Brien, M. 1979. *The idea of the American South, 1929–1941*. Baltimore: Johns Hopkins University Press.
Park, R. 1924. *Letter to R. D. McKenzie dated 2 January*. Park Papers, Hughes Collection, Cambridge, Massachusetts.
Park, R. 1926. The urban community as a spatial pattern and a moral order. In *The urban community*, E. W. Burgess (ed.). Chicago: University of Chicago Press.
Park, R. 1937. *Letter to Harry E. Moore dated 18 May*. Odum Papers, Southern Historical Collection, University of North Carolina Library, Chapel Hill, North Carolina.
Reed, J. 1982. *One South: an ethnic approach to regional culture*. Baton Rouge: Louisiana State University Press.
Relph, E. 1981. *Rational landscapes and humanistic geography*. Totowa, NJ: Barnes & Noble.

Royce, J. 1908. Provincialism. In *Race questions, provincialism and other American problems*, J. Royce (ed.). New York: Macmillan.

Sack, R. D. 1974. Chorology and spatial analysis. *Annals of the Association of American Geographers* **64**, pp. 439–52.

Sack, R. D. 1980. *Conceptions of space in social thought: a geographic perspective.* Minneapolis: University of Minneapolis Press.

Sauer, C. O. 1938. *Letter to Frank Aydelotte dated 6 June.* Sauer Papers, Bancroft Library, University of California, Berkeley.

Schaefer, F. K. 1953. Exceptionalism in Human Geography: A Methodological Examination. *Annals of the Association of American Geographers* **43**, pp. 226–49.

Singal, D. J. 1982. *The war within: from Victorian to modernist thought in the South, 1919–1945.* Chapel Hill: University of North Carolina Press.

Smith, N. 1984. *Uneven development.* Oxford: Basil Blackwell.

Sopher, D. 1972. Place and location: notes on the spatial patterning of culture. *Social Science Quarterly* **53**, pp. 321–37.

Steiner, M. 1979. The significance of Turner's sectional thesis. *The Western Historical Quarterly* **10**, pp. 437–66.

Suttles, G. D. 1984. The cumulative texture of local urban culture. *American Journal of Sociology* **90**, pp. 283–304.

Tobey, R. C. 1981. *Saving the prairies: the life cycle of the founding school of American plant ecology.* Berkeley: University of California Press.

Turner, F. J. 1959. *The significance of sections in American history.* Gloucester, MA: Peter Smith.

Wallerstein, I. 1974. *The modern world-system: capitalist agriculture and the origins of the European world-economy in the 16th century.* New York: Academic Press.

Wallerstein, I. 1979. *The capitalist world-economy.* Cambridge: Cambridge University Press.

Weber, M. 1949. *The methodology of the social sciences.* Translated by E. Shils & H. Finch. New York: Free Press.

Whitehead, A. N. 1957. *Science and the modern world: Lowell lectures, 1925.* New York: Macmillan.

Wirth, L. 1965. The limits of regionalism. In *Regionalism in America*, Merrill Jensen (ed.). Madison: University of Wisconsin Press.

4

Modernism, post-modernism and the struggle for place

DAVID LEY

"A place for everything, and everything in its place," wrote Samuel Smiles, the "quintessential Victorian", in 1876 (Kern 1983, p. 182). But the optimistic patterning of middle-class Victorian culture, its tidiness and propriety, were already experiencing dislocation which would soon wrench its ordered life asunder, and introduce that peculiar modern sensibility recently explored by Marshall Berman (1982) where "all that is solid melts into air." Among the first and most disorienting of assaults was the frontal attack on religious belief. Only a few years after the text of Samuel Smiles, Nietzsche's madman declared that "God is dead", though the consequences of disbelief were not lost to him: "What were we doing when we unchained this earth from its sun? Whither is it moving now? Whither are we moving? Away from all suns? Are we not plunging continually? Backward, sideward, forward, in all directions? Is there still any up or down? Are we not straying as through an infinite nothing? Do we not feel the breath of empty space?" (Kern 1983, pp. 178-9).

From the stability of an ordered place to the vertigo of empty space: this is a peculiarly geographic metaphor, and I shall touch briefly on its implications for the discipline of geography later. But in this chapter I have a broader objective, to trace the relations of place and space to the discourse of modernity over the past 100 years. Of course this discourse represents more than the history of ideas; it expresses in addition the history of the built environment, the projection of modern spatial theory onto the landscape. Again, however, we cannot stop here, with a seemingly irrepressible world spirit that flows through and over space, transforming the landscape into its own image – though such Hegelianism is certainly pervasive in the confident evolutionary slogans of many of the pioneers of modern architecture.[1] Both the discourse and the practice of modernity are inherently political; they evoke causes, mobilize interests, and frequently, especially in periods of transition, they galvanize conflict. There has invariably been a struggle for the definition and the making of the built environment, a struggle to empty out and purify space in the early campaigns of the modernists, and an equally keen struggle to once again fill space with meaningful references in the often muddled project of post-modernism. In this chapter I shall examine several themes in the struggle for place; first, in the design professions, more briefly in the

44

discipline of geography, and not least in political practice in the streets. Beginning with the late 19th-century attack upon bourgeois culture by the early modernists, we shall consider also current post-modern criticism of modern landscapes. We shall see too that the struggle to define the language of space and place has often formed part of a larger struggle for the definition of culture itself.

The challenge to bourgeois culture, 1880–1920

Several studies have identified Vienna as a significant centre in the turn of the century struggle for modern ideals and practices against an entrenched bourgeois culture, a conflict which extended over many intellectual and artistic genres, including the making of the built environment. In his masterly interpretation of the grand ceremonial buildings and public spaces of Vienna's Ringstrasse, constructed in the generation following 1860, Carl Schorske reads the landscape as "a cultural self-projection...an iconographic index to the mind of ascendant Austrian liberalism" (Schorske 1981, pp. 26-7). Yet for all its success, the Ringstrasse was not without its critics, including Otto Wagner, a "rational functionalist" and pioneer in modern architecture, and Camillo Sitte, a "romantic archaist" and advocate of the picturesque, of historic not modern styles, and of human scale spaces recreating the experience of *gemeinschaft* folk community (ibid, p. 13). For now, the reaction of Wagner is of more interest, not least because his criticisms anticipated the direction urban design would follow during the next three generations. Wagner was unimpressed by the classicism of the Ringstrasse which seemed a false historical metaphor for the age of iron and steel. So too the fussy clutter of the Ringstrasse's ornamentation compared poorly with the clean lines, the honest relationship between form and function exemplified in the industrial machine, a truer metaphor for the contemporary age. Wagner was not of course an isolated critic of bourgeois culture in *fin-de-siècle* Vienna. We are told that the search for artistic integrity, for authentic forms of expression, the problems of language and communication, all were a virtual obsession among artists and intellectuals who despised the pretence and superficiality they perceived in the last decades of the Hapsburg Empire, an inauthenticity which denied the political and economic realities of the times (Janik & Toulmin 1973).

The excesses of bourgeois "good taste" appeared in domestic as well as in public design and architecture, as is evident from the following description of the homes of the Viennese middle class (cited in Janik & Toulmin 1973, p. 97):

> Theirs were not living-rooms, but pawnshops and curiosity shops. . . .(There was) a craze for totally meaningless articles of decoration...a craze for stain-like surfaces- for silk, satin and shining leather; for gilt frames, gilt stucco, and gilt edges; for tortoise shell, ivory, and mother-of-pearl, as also for totally meaningless articles of decoration, such as Rococo mirrors in several pieces, multi-colored Venetian glass,

fat-bellied old German pots, a skin rug on the floor complete with terrifying jaws, and in the hall a life-sized wooden Negro.

It was an escape of masquerade: "The butter knife is a Turkish dagger, the ash tray a Prussian helmet, the umbrella stand a knight in armor, and the thermometer a pistol." Such cultural aberration and its critics were not peculiar to Central Europe. In Britain, writers such as George Bernard Shaw and Oscar Wilde engaged with gusto in the demasking of bourgeois charades; *The Importance of Being Earnest* is a title containing a pun directed not only at a Victorian character but also at a Victorian disposition. Moreover, in the United States as well, the young Frank Lloyd Wright could find houses in 1894 which sounded as if they were replicas of middle class Vienna: "Too many houses, when they are not little stage settings or scene painting, are mere notion stores, bazaars or junk-shops" (Wright 1975, p. 235). And a generation later Le Corbusier (1927, pp. 18, 91) railed against house decoration as "a conglomeration of useless and disparate objects...the intolerable witnesses to a dead spirit."[2]

There is some evidence that the hoarding of esoteric objects by the middle class did indeed conceal a deeper insecurity, for in addition to the polemical assaults from without, middle class culture was also faltering from within. In Vienna, the expressionist painter Kokoschka explored (like Freud, in another genre) the inner confusion he perceived behind cultivated exteriors: "the people lived in security, yet they were all afraid. I felt this through their cultivated form of living, which was still derived from the baroque; I painted them in their anxiety and pain" (Janik & Toulmin 1973, p. 101). Part of the cultural malaise accompanied the new uncertainties of religious belief in an age of growing scepticism, in the wake of Darwin and the practical and philosophical materialism of the myth of progress. "It is so new to me" wrote John Ruskin "to do everything expecting only death" and the theme of spiritual dissolution runs broadly through British intellectual life of the period (Miller 1963; Barnard 1984). By the time of Yeats, the battle had been lost, the integrative values of middle class culture had come unstuck: "Things fall apart, the centre cannot hold."[3] The process of cultural dissolution among the American middle-class has been traced in some detail after 1880, as rapid economic and technological change, urbanization and the erosion of traditional beliefs conspired in complex ways to produce a disorientation of personality and culture: "As supernatural beliefs waned, ethical convictions grew more supple; experience lost gravity and began to seem 'weightless' (Lears 1983). Once again we encounter the vertigo of empty space.

A range of disparate yet related movements which peaked in the 1890s sought to redefine culture: the arts and crafts movement, symbolism, the aesthetic movement, art nouveau, the Vienna Secession – each sought to establish a new integrity in high culture, in which there was still room for history and regional tradition, for spiritual and emotional expression, for recognizable representations of beauty, for decoration to take an authentic form. Yet these movements were by and large short-lived, and served primarily to provide a transition to a new and quite different earnestness, in the pursuit of a serious utilitarian rationalism, the spirit of the modern movement, which substantially removed the intangible, the metaphysical, even (or so it seemed) culture itself in favour of an objective and functional logic, the spirit of progress, a spirit of "sincerity and purity" in its relations to the modern era.[4] That logic, I shall argue, created spaces not places, masses not meanings, and posed in a new and urgent way the problem which remains with us today, the problem of conceptualizing, designing, and building meaningful places. Not the least intractable issue to emerge, as we shall see, is the problematic status of contemporary culture itself.

The modern movement and culture as production

Implicit in our discussion is the tight bonding between the arts, including architecture, and society, a mutuality of text and context which should not be oversimplified (LaCapra 1983). This relationship bore a radical message, for "the decadence of taste and the arts was a direct consequence of the decadence of the society of which they were an expression" (Branzi 1984, p. 13). Indeed part of the critique of middle class taste and culture was directed at its inauthenticity in speaking to the modern world, its perceived dishonesty in refusing such a dialogue. Art then that was true to its time could not help but be adversarial and subversive of art, cultural taste and social life which denied such an engagement. The contours of such opposition were made clear by Adolf Loos, a Viennese architect and culture critic, who in *Ornament and crime* in 1908 developed his attack upon the continuing privileged status of ornament in the work of the reformers of the Vienna Secession, a status he saw as regressive if not criminal (Janik and Toulmin, 1973, Frampton 1985). Ten years earlier he had satirized the Ringstrasse for its celebration of patrician values which did not represent the realities of urban society in Vienna. Now the argument became more radical; middle-class life, neither patrician nor peasant, had no real tradition and thus *no real culture* to celebrate. Far better then that design be purely objective and functional, purely utilitarian: "the meaning" he wrote, "is the *use*" (Janik & Toulmin 1973, p. 207). Juxtaposed to more conventional designs, Loos' buildings assaulted the values of mainstream culture. He built a simple, unadorned structure opposite the Imperial Palace in Vienna, a building whose "very simplicity and functionality were regarded as an intentional insult to the Emperor" (ibid, p. 100). In the struggle for space, the fine arts had a subversive role in challenging also the authority of the state.

Meaning in objects, in the built environment, was defined in utilitarian terms, so that a functional and uniform aesthetic became the appropriate expression of a machine-based mass society. Uniformity became a universal. "I propose one single building for all nations and climates" Le Corbusier was to declare later (Brolin 1976, p. 44), and one can see the prototype of this modern landscape in the bare, unadorned and geometric structures built by Wagner and Loos in Vienna during the first decade of the 20th century. These were structures born of a simple geometry, a universal logic devoid of historical and regional references, references which were in any case obsolete if Loos was right that modern urban society had no authentic traditions left to preserve. The cultural agnosticism of the modern movement in architecture has continued to the present. Like Loos, the late modern Canadian architect Arthur Erickson can see no authentic culture in industrial society; the built environment of the modern city is "a profound charade that disguises momentarily the emptiness of our souls." However, around this zero point, Erickson like the early modernists, sees an opportunity for universalism: "I am fortunate that I can stand in Canada, a country without a culture, and look at the world" (Iglauer 1981, p. 20, 58).

Much of this argument may be summarised if we return for a final time to Vienna and consider the house built by Wittgenstein for his family in the 1920s. It expresses the same ideals of honesty, simplicity, precision and a self-referential logic that characterized the works of critical Viennese intellectuals during this period. These same ideals were also taken up by the Circle of philosophers and social scientists in Vienna who were simultaneously striving to formulate a universal philosophy of science, a philosophy that systematically excluded from consideration the intangible and the metaphysical, the realm of consciousness, spirituality, values, and culture, as unknowable. Wittgenstein, author of the influential *Tractatus Logico-Philosophicus*, created a built form that was a tight synthesis of architectural modernism and logical positivism. His biographer detects the same aesthetic in the abstractions of both his philosophy and his house design: "(the house's) beauty is of the same simple and static kind that belongs to the sentences of the Tractatus" (Von Wright 1984, p. 11). The house presents a remarkable projection of intellectual purity. It strives after complete honesty in its rejection of decoration, carpets, or any distinctively regional or cultural content to obscure the integrity of the building materials and a consistent cubist geometry. It was constructed to demanding technical qualification, requiring a level of mathematical precision that local engineers often had great difficulty in meeting. From the bare light bulbs, to the unpainted walls, to the lack of human proportions, there was in their "house turned logic" as Wittgenstein's sister described it (Leitner 1976, p. 23), a complete expression both of the uncompromising abstractions of the Vienna Circle, and of the ideology of modern architecture, for whom after Le Corbusier, a house was "a machine for living in" and a street "a factory for producing traffic." The modern architecture which elaborated this design vocabulary over the following decades did not stray far from such adherence and admiration for the machine metaphor. Arthur Erickson found in North American urban cores around 1970 "a desolating impersonality as if no creature but a machine had made them." But just as we anticipate a critical rejoinder, Erickson surprises us with a startling affirmation: "But this was also their vitality...purely utilitarian space" (Iglauer 1981, p. 93). Within this somewhat cavalier justification for the machine-age aesthetic in urban planning and design was contained the basis for a response which was to develop with increasing fury during the 1960s and 1970s.

If Loos accomplished the assault on bourgeois culture, it was left to Le Corbusier to make explicit the topography of a new culture truer to the circumstances of the 20th century.[5] This was an engineering or machine age aesthetic, creating a built environment where modern man "*can* be proud of having a house as serviceable as a typewriter" (Le Corbusier 1927, p. 241). A blueprint he completed with his cousin for an economy car for mass production, and a standardized house design named Maison Citrohan, a pun on the automobile manufacturer, locate Le Corbusier unambiguously in the technological thrall of an early 20th-century society fascinated with the capacity of rationalism in all its forms to propel society forward into a brave new world. Almost a quarter of his book, *Towards a new architecture*, is given over to the admiration of steamships, airplanes and automobiles: "The steamship," he noted, "is the first stage in the realization of a world organized according to the new spirit" (ibid, p. 103). This required a new language, a transition from "the elementary satisfactions" of decoration to "the higher satisfactions" of mathematics (ibid, p. 139). From this viewpoint, Le Corbusier regarded the primitive geometries of cubist painting as attaining "attunement with the epoch" for "geometry is the language of man" (ibid, p. 19, 72). In the same way that space in modern science was defined relative to a moving reference point, so the cubists broke up objects to examine them simultaneously from several vantage points. And so too the geometric slabs of Bauhaus design expressed a "new conception of space, with its urge toward freely hovering parts and surfaces" (Giedion 1967, pp. 484–5; also Kern 1983, Ch 6). Suspended and intersecting concrete slabs, simultaneously inside and outside buildings bore unmistakable affinities with cubist space. And they were built to be seen in motion, by mobile spectators passing by on freeways, or peering down from an airplane.[6]

The project of the Bauhaus was the integration of art and industrial life, and it has been noted that cubism as an art and design form implied "a deep interiorization of the logic inherent in assembly-line production" (Branzi 1984, p. 18). For this design programme, the factory, already a dominant economic and social institution, became the source of cultural and artistic inspiration as well. A number of the important early commissions of the modernists were for factories. Indeed, in contrast to the existing house form, an "old and hostile environment", an "old coach full of tuberculosis", the profiles of grain elevators and factories, geometric, minimal, functional were altogether "the magnificent first-fruits of the new age" (Le Corbusier 1927, p. 31, 277, 284).[7] "A modern building," wrote Gropius "must be true to itself, logically transparent and virginal of lies or trivialities, as befits a direct affirmation of our contemporary world of mechanization and rapid transit" (Brolin 1976, p. 51). One senses in these paeans, a Hegelian synthesis where the machine is the incarnation of geist, the world spirit of the new age. In a society heavily impregnated by Darwin, the modernists were confident that when it came to an evolutionary imperative, they had, if not God, then chance, firmly on their side.

Mass production required a mass culture, just as a mass culture required mass production. The machine age was seen as liberating, for through standardization mass production alone could address the housing problem in large industrial cities; a housing assembly-line offered the only logical solution to providing the material basis for a new social order.[8] And beyond this, standardized products supported an egalitarian culture. Here was the utopian political edge of modern architecture, the society implicated in the design solution. In a slightly different form, Frank Lloyd Wright carried forward the emancipatory potential of the engineering aesthetic: the machine, he wrote in 1927, is "the ideal agent of Usonian democracy" (Wright 1975, p. 135).

It was this utopianism which led Le Corbusier, Wright and other prophets of modernism to integrate architecture with planning. Through simple repetition, with some consideration of density and (always) transportation needs, the design solution for the individual household was extrapolated into a series of urban plans: the city of towers (1920), the Ville Contemporaine (1922), the Ville Radieuse (1930) from Le Corbusier, Broadacre City (1932) from Wright, each of them egalitarian, repetitive and functional (compare Gold 1985). In Le Corbusier's case they were invariably massive. The city of towers, apartment blocks up to 700 feet high and 500–600 feet deep, housing up to 40,000 people, was according to Le Corbusier, "a reasonable idea", "an architecture worthy of our time" (Le Corbusier 1927, p. 58).

Characteristics of the engineering aesthetic which included efficiency, functionalism and impersonality were shared by a broader interwar movement in urban planning. A 1923 article entitled "Reasons for Town Planning", published symptomatically in the *Canadian Engineer*, laid out the successful formula: "Good city planning is not primarily a matter of aesthetics, but of economics. Its basic principle is to increase the working efficiency of the city" (van Nus 1979). Inevitably the appropriate model for urban administration was perceived to be the rational business corporation. For Le Corbusier (1927, p. 284) "big business is today a healthy and moral organism", and other contemporaries agreed that the business corporation could "introduce into city government the standardization and scientific management already found in industry" (Zorbaugh 1929, p. 272). This latter statement, by a Chicago urban sociologist, suggests that the influential interwar Chicago School was nudged toward the centralized model of management and control of urban life. Like Adolf Loos, though perhaps for different reasons, the Chicago sociologists saw everywhere social disorganization, the dissolution of local folk cultures in the industrial city. The source of a new moral order must arise out of the new conditions; it was futile to attempt to rebuild local urban cultures. What was this new moral order? Le Corbusier would no doubt have been satisfied by the value-laden distinction drawn by the Chicago sociologist Louis Wirth between industrial and agrarian society. "There is a city mentality which is clearly differentiated from the rural mind. The city man thinks in mechanistic terms, in rational terms, while the rustic thinks in naturalistic magical terms" (Wirth 1925; also Park 1925). The machine or magic! These were the provocative metaphors separating the modern from the pre-modern mind, demonstrating once again the deep penetration of the ideology of industrial production into domains of culture and even psychology. With unwavering consistency, Wirth (like Zorbaugh) was drawn to the model of the business corporation in urban administration precisely because the purely rational corporation "has no soul" (Wirth 1964).

It is worth emphasizing that these interpreters of the city did not regard their proposals as arbitrary impositions. The cauterizing of interpersonal relations implicit in the bureaucratic model simply institutionalized what Wirth and Simmel had called the blasé attitude, the emotional neutrality which "naturally" governed so much social interaction within the city. But the rational attitude had to be cultivated, it required its professional translators to make the new order transparent to a mass public; in the arts this was the self-appointed mandate of the avant-garde, in the city it was the task of a new group of social engineers, including the planner, the social worker, the traffic engineer, and the city manager, to inculcate the spirit of rationality and professional disinterest. This rational disposition provided a source of theoretical knowledge perceived to be superior, and both the architectural modernists and the municipal reformers in Chicago held a sceptical if not disparaging view of the folk knowledge of the public.[9] Giedion's view of the architect's constituency as knowledgeable agents was little short of contemptuous: "Of course, the (architectural) problems that have to be solved are not posed by any conscious expression of the masses. For many reasons their conscious mind is always ready to say "No" to new artistic experiences" (Giedion 1967, p. 598). Gropius found his working-class clients too "intellectually undeveloped" to be consulted in his mass housing projects, while Le Corbusier also sensed the need for people to be "re-educated" to interact meaningfully with his urban vision (Knox 1987). It should come as little surprise that for Giedion, a student of Le Corbusier, Robert Moses, the grand builder of New York's highways and public works for 40 years, represented the climax of modernity applied to the city. For Moses also, city-building was a technical problem of clearance and construction: "more houses in the way...more people in the way – that's all...When you operate in an overbuilt metropolis, you have to hack your way with a meat ax" (cited in Berman 1982, pp. 193-4).

The post-modern challenge: toward sensitive urban place-making

It is not possible in this chapter to itemize in any detail the patterns which accompanied this ideology, but in its major components it may be readily recognized: the high rise, freeway landscape of the urban core with the standardized suburban tract home, a collage of the visions of Le Corbusier and Frank Lloyd Wright, fought for by functionaries like Robert Moses and fuelled by development capital and government bonds. But by the 1960s, the promises of the prophets of the 1920s had grown increasingly hollow. A corporate urban landscape, the product of an increasingly corporate society, became the legacy of the modern movement, and through the 1960s and 1970s a critique emerged that the planning and design of the modern city was a blueprint for placelessness, for anonymous, impersonal spaces, massive structures, and automobile throughways. Interestingly this is precisely what contemporary modernists (like Arthur Erickson) admire, and what the early modernists intended in their desire to build in conformity with the spirit of the age. Sixty years later, however, the consequences for urban places are clear. Looking at the Texas urban skylines which, as the most influential American practitioner of modern architecture, he helped to build, Philip Johnson recently reflected that "There is no sense of place. These skyscrapers are like individual tombstones in a cemetary."[10]

Johnson's verdict is strikingly reminiscent of Max Weber's pessimistic judgement on modernity, for the cultural agnosticism of the pioneers of modern architecture was itself one instance of the broader disenchantment of the modern mind, and its consequence, a numbing "iron cage" which forecloses so much human creativity. With the growth of science and technology, observed Weber, people have lost their sense of the sacred: "Reality has become dreary, flat and utilitarian, leaving a great void in the souls of men which they seek to fill by furious activity and through various devices and substitutes" (cited in Bell 1980, p. 327). The metaphor of the void is remarkably persistent in accounts of 20th-century culture: in early examples of the modern novel (Kern 1983); in the "hollow men" of T.S. Eliot and others; "the void within" is described as a modern personality trait by Christopher Lasch (1978); while the philosopher William Barrett describes how "A huge gaping void has opened up...(with)...an unconscious aching to fill this void, to see meaning in the universe" (cited in Agena 1983). The Japanese modern architect, Arata Isozaki explains his inert design grids in terms of the agnosticism of the present age, as "a metaphor of degree zero, or a void at the centre" (Jencks 1980, p. 107). What Weber lamented, modern architecture has seen as necessary and indeed normative.

The critical stance of Weber has been carried forward by the theorists of the Frankfurt School, and notably in the present era by Jurgen Habermas in his critique of the penetration of everyday life by the instrumental rationality of the state and the market (Bernstein 1985a). Habermas (like Berman but unlike Weber and other authors) sees resistance as built into the very fabric of modernity.[11] In particular for critical theorists, art and aesthetics provide an important potential source of resistance (Bernstein 1985b, Jay 1985). As a result it is necessary for Habermas (1983) to reject Daniel Bell's (1976) thesis that the corrosive edge of modernity lies within the arts themselves. Rather the arts offer the possibility of a post-modernism of resistance, which has been counterposed to a conservative post-modernism of reaction (Foster 1983).

Certainly, the modern movement's rejection of history, meaning and the transcendent, its reduction of culture to the language of production, has provoked an important counter-current in the past 20 years. "Reality, it turns out, is not like a machine after all" writes Kathleen Agena (1983) in an informative short paper documenting the contemporary "return of enchantment" across a broad spectrum of popular and elite culture. A philosophical reorientation has emerged in the arts and literature, in architecture and planning, and also in the social sciences and has been expressed politically in the city by neighborhood activism and social movements engaged in a struggle to preserve and enhance places that mattered. A critical and often adversarial literature advocated the reconstitution of meaning, a new respect for subjective needs, the rediscovery of cultural symbols in the built environment. In planning and architecture, criticism – sometimes satirical, such as the "world of absolute madness" in the design extravaganzas of the radical groups Archizoom and Superstudio (Branzi 1984) – was directed against the decentering of a functionalist landscape away from human sensibilities and human proportions, the placelessness which seemed to accompany centralized corporate decision-making and standardization in a mass society (Relph 1976; Jencks 1981; Boyer 1983). Against the uniformity of the modern movement, arose a renewed interest in the specificity of regional and historical styles, and the diversity (not the uniformity) of urban subcultures.

In urban design the transition from a cubist grid to what geographers have called a sense of place reveals a more complex attention to theories of space in the plural styles of what has become known as post-modern architecture. For some architectural theorists, the philosophical inspiration is phenomenology, for others it may be semiotics, but in each instance the objective is the construction of forms which suggest and evoke symbolic associations, "sensitive urban place-making" (Jencks 1981, p. 82). In contrast to the isotropic space of modernism, post-modern space aims to be historically specific, rooted in cultural, often vernacular, style conventions, and often unpredictable in the relation of parts to the whole. In reaction to the large scale of the modern movement, it attempts to create smaller units, seeks to break down a corporate society to urban villages, and maintain historical associations through renovation and recycling. Jencks sees in post-modern design "the return of the missing body", an attempt to restore meaning, rootedness and human proportions to place in an era dominated by depersonalizing bulk and standardization: "Surely the ultimate paradox and strength of Post-Modernism is its adamant refusal to give up the imperatives of the spirit at a time when all systems of spiritual expression have been cast into doubt" (Jencks & Chaitkin 1982, p. 217). The post-modern project is the re-enchantment of the built environment.

Symbols in a cultural void

But the project immediately confronts a formidable historical obstacle, which casts in doubt its emancipatory promise. With what symbols are the post-modern magicians going to make their magic? And then there is the problem of pluralism, the fragmentation of modern societies into often factious taste cultures (Scott Brown 1980). Is there any hope of a consensus in such pluralism? No wonder then that before such a daunting task, the characteristically post-modern "crisis in content" impedes contemporary design initiatives. For as Jencks (1983, 1985a) puts it (after yet another reference to the void at the center of society) in our agnostic society what do we have to symbolize? The so-called "death of God" removed the possibility of religious symbolism. Other historically transcendent symbols such as the liberal belief in progress or the nation state are similarly tarnished. Indeed the decline of "meta-narratives" is seen as part and parcel of the post-modern condition (Lyotard 1984). The interior designer, Andrea Branzi describes the New Design, a contemporary movement sympathetic to cultural expression, as including "new inspirations deriving from irony, curiosity, surprise and friendliness" (Branzi 1984, p. 148). Laudable though these sentiments may be, and they are pervasive in post-modern design, do they represent the limit of the search for transcendent symbols today? Do they meet the worthy requirements of the pioneers of the modern movement for authentic content in design?

This is precisely the conundrum recognized by Camillo Sitte as he confronted the nascent modernism of 19th-century Vienna. Where in the fragmentation of modern life could one find an integrating myth? Sitte found it in a romantic and nostalgic nationalism which sustained the myth of a medieval communitarian society (Collins & Collins 1965, Schorske 1981, p. 69). In the decade of Rambo and the Falkland Islands his nationalist solution has a current ring. So too the most successful contemporary solution to the problem of architectural content, particularly in popular terms, is the past, a historicism of conservation and revival of vernacular forms. But for all their familiarity, regional sensibility, human scale and market success, one has to wonder about the lack of confidence in symbolizing current cultural experiences suggested by the vernacular style. There is the same disquiet in popular culture, where Fredric Jameson (1983) has questioned the ubiquitous genre of nostalgia movies "as though we have become incapable of achieving aesthetic representations of our own current experience." Moreover, particularly (though not exclusively) in North America historicism may consist of a giddy eclecticism, a mixing and matching of the design elements of a range of historic styles, where for example a fashionable townhouse complex called The San Franciscan might include Greek columns, Moorish lattice and a triumphal arch. Here we may not be far from the style excesses condemned by Adolf Loos and Frank Lloyd Wright at the turn of the century, and their message becomes, perhaps, once again topical in such faithful restorations as the 1894 Union Station in St Louis with its riot of styles and colors (Time 1985).

A second post-modern theme is self-consciously populist. It abandons the elitism of modern design theory and throws itself headlong into the brash stage set of the neon commercial highway, of which the main strip in Las Vegas must surely be the ideal type (Venturi, *et al.* 1972). This is a bold solution theoretically and architecturally in its dialogue with everyday popular culture which is roundly condemned by those with "taste", but inevitably it invites the criticism of kitsch in both its modest structures and its landmarks, such as the celebrated symbol of a Chippendale chair (or is it a coin machine?) atop New York's A T & T Building, the glass pastiche of the Houses of Parliament in Pittsburgh, or the design jokes and parodies by the pop art group, SITE projects, in suburban shopping centres. Such design may be humorous, sometimes ironic and thus critical, but rarely does it seem more than an optional gift wrapping to the surface of the built environment. It titillates and teases, but risks dismissal as inauthentical froth, the celebration only of a society that is skin deep.[13]

A third and novel solution had been created by Charles Jencks (1985b) in the interior design of his own townhouse in London. Observing that "symbolic architecture fulfills a desire", he poses the question, but what "remains worthy of celebration in a secular age?" His answer is nature, and his house displays a consistent and self-conscious program of natural and cosmological iconography. For example, four rooms representing each of the seasons revolve around the central Sun Stair. In a perfection of detail, the Sun Stair itself has 52 steps and three rails (sun, earth and moon) that rotate upwards in a spiral motion. It is a bold and creative, if somewhat self-conscious experiment, but its naturalism belies a rather lonely universe, suggesting that we modern beings have very little we can be sure of.

So while its questions are pertinent, there is limited indication that post-modern design has yet arrived at fully convincing cultural solutions. Moreover, an experimental proposal by Kenneth Frampton (1983) to establish a critical regionalism in design, influenced by the critical theorists, has proven no more successful. The emphasis on facade appearances in post-modernism raises the possibility that solutions are primarily aesthetic. The sensuous nature of post-modern townscapes, the ominous reference by interpretors to parallels with a film set, pose the question of the interests evoking post-modern magic. The concentration of post-modern codes in North America among clusters of up-market townhouses and leisure resorts for the upper middle class provides an answer of sorts. Is the magic of post-modernism now part of the magic of the sign, part of the cultural codes of consumption (Wernick 1984, Ley 1985a)? If so, then post-modern theory has befallen the fate of other cultural symbols where a rebellious threat has been converted into a marketable commodity, which as a lifestyle sign, incorporates its users and sellers harmlessly back into society even while giving an impression of their separation from it.[14] So while post-modernism is convincing as critique, its design solutions as yet fall short of the ambitious emancipatory program laid out for it, and already there are rustlings of a neo-modernism. The struggle in design is another example of the fragile status of a contemporary post-modernism of resistance; as Habermas (1983) acknowledges, "the chances for this today are not very good."

Planning theory and the struggle for neighborhood

The post-modern struggle for place has also been fought in planning and urban politics over the control of streets and neighborhoods. Here we leave the ideology of Otto Wagner's Vienna and turn to Camillo Sitte, the critic of Vienna's modern landscapes, whose "communitarian vision of re-humanized urban space" set an important precedent for a major present-day communitarian, Jane Jacobs (Schorske 1981, pp. 25, 72).

For Le Corbusier and the early modernists, the street was "a factory for producing traffic", and as the corporate city took shape its rational plans led to the demolition of neighborhood upon neighborhood in the interests of freeway construction, urban renewal, and various public works. Marshall Berman (1982) tells the story of the destruction of his own childhood neighborhood in the Bronx during the late 1950s and early 1960s as Robert Moses drove the Cross-Bronx Expressway through the center of the borough, displacing some 60,000 people. Moses' metaphor of wielding the urban meat ax, his respect for an ethereal public good, but disdain for people, reveals a dehumanizing capacity in modernity. The ontology of a mass democracy overlooked the existence and the needs of individuals and minorities. In 1924 Mies Van der Rohe observed that "The individual is losing significance; his destiny is no longer what interests us" (Watkin 1977, p. 38). But the city comprises individual households, individual streets, and individual neighborhoods, and during the 1960s a range of authors and activists revived the humanist protest of Kierkegaard, to protect the integrity of the individual against the tyranny of the anonymous masses. The task was threefold: an ontological challenge directed against the abstraction of mass society or the public good; an epistemological challenge against the instrumental rationality of the social engineers who were its functionaries; and a political challenge against the power elite who shared their visions.

Perhaps the most celebrated of the early critics of modern planning was Jane Jacobs, who sought to demonstrate the destructive effects of centralized rational planning upon urban life. Her influential book, which she described as an "attack on current city planning and rebuilding" (Jacobs 1964, p. 16), identified a faulty logic in detached technical plans. To a remarkable degree, her comprehensive program, so innovative in 1960, was adopted by planning departments in the 1970s. What was at stake, Jacobs argued, was the preservation of community and the personalizing of space. "People places", vital, animated, settings were being destroyed by detached, rational planning; the answer was to plan at a human scale, to build an environment which in its landmarks, folk allusions, and meeting places would sustain intersubjective associations and a sense of place. In opposition to the segregated land uses of existing plans, diversity of people and land uses was a cardinal principle: "Planning for vitality must stimulate and catalyze the greatest possible range and quantity of diversity among uses and among people throughout each district of a big city (Jacobs 1964, p. 408). This conclusion was reached in large measure from her everyday experience of her own street in New York, the daily performance of a Hudson Street "ballet", the comings and goings of neighbors and strangers, whose diversity and varied time–space rhythms defined a distinctive "ecology and phenomenology of the sidewalks" (Berman 1982, p. 322). This radical penetration of everyday life into a planning prospectus formed the focus of her critique of modernism. Ontologically it revealed the existence of a meeting of people and place that had been obscured from the detached decision-makers; epistemologically it showed a new appreciation of the strength of folk and personal knowledge; and politically it advocated a participatory method which established direct communication between everyday life and planning practice.

As is well known, the struggle to preserve cherished places was carried by neighborhood groups into the streets and council chambers of major cities for a decade or more. There were some important successes: expressway plans were torn up, urban renewal was arrested and redefined, heritage structures were preserved, neighborhoods were downzoned, and through traffic diversion, landscaping and greater pedestrian access, livable streets began to be recovered (Appleyard 1981). One by one planning goals set out by Jacobs were adopted and the initiatives of the corporate city were resisted (e.g. Lorimer 1983). It seemed indeed that the lifeworld was reclaiming for itself some of the territory invaded by private and public corporatism.

But by the 1980s the gains seem less secure. What Berman calls the "undertow of nostalgia" in Jacobs' thinking shows, like post-modern architecture, a certain failure of confidence in engaging the present. Jacobs writes as a "radical conservative" (a phrase associated with John Ruskin, whose symbolist and somewhat romantic worldview has many points of contact with post-modernism)[15] and as the 1980s progress the potentially conservative content in this program is unfolding. The idealization of districts like Hudson Street has had the effect of changing them into commodities for which the upper middle class will pay a substantial entry fee. Initial entrants into these districts commonly have left-liberal sympathies and endorse the sentiment of social difference, but as the process unfolds and prices inflate a more conservative ethos often takes hold (Ley 1985b). Throughout the period the steady displacement of tenants and lower income groups has been underway, and the end of the process sees the creation of a gilded ghetto which no longer exudes social and cultural diversity. In short, the consequences of neighborhood preservation in the inner city have often displayed an unintended elitism. In the words of a frustrated Canadian planner: "Our strategies are retaining and improving communities for a few higher income households" (McAfee 1983). Moreover as these districts consolidate their advantage through sympathetic aldermen and revised zoning schedules, so they become resistant to the construction of social housing and other incursions of the welfare state.

The substitution of individual for communal goals reveals an ambivalence which Berman recognizes in his own relation to neighborhood. Though a participant in the critical social movements of the 1960s, he is drawn to the realization that despite his lament for the destruction of his childhood home in the Bronx, he too would have abandoned the district and its familial and communal memories once social mobility enabled him to move elsewhere. This paradox, an ambivalence of intent and consequence, runs deeply through post-modernism. The post-modern project is particularly the project of the cultural new class, representatives of the arts and the soft professions who came to political awareness in the 1960s and were receptive to the oppositional ideals of the counter-culture (Martin 1981, Edgar 1986). But through the 1970s and 1980s hippies have all too readily become yuppies, as the subjective philosophies of phenomenology and existentialism which opposed the impersonality of modernism in the 1960s and redeemed the individual have been directed inwards, and the celebration of meaning has often shifted subtly to aestheticism, and the celebration of the meaning of the self (Lasch 1978).

There are indications of the same shift in post-modern literary theory, where the post-modern venture had its roots in part in phenomenology and existentialism, was invigorated by the oppositional movements of the 1960s, and like post-modern architecture used to special effect the tool of irony (Arac 1986). But in its important engagement with subjectivity, the presence or absence of the author can become an unsatisfactory game of literary hide and seek, as the writer leaves tantalizing clues for the reader to guess his whereabouts, and the text itself is left suspended until the game is complete.

There are other expressions of the same dynamic. In 1968 the reclamation of the streets and the proclamation of "exhilarating, joyous festival" was an important theme of the Paris student riots (Poster 1975, p. 373). But for the next six years street activity of any kind was discouraged by the police, and the modernist commitment to high rises, expressways, and large public works returned, promoted by a neo-Gaullist President (De Lacy 1983). In the late 1970s a new initiative arose "to make Paris gay again" which included the reanimation of street life, encouraged by pedestrianization, landscaping, and cultural happenings in public places. Interestingly, this initiative to turn Paris into a "fun city" has been led by a right-wing mayor who well understands that the aestheticism of his constituency does not require progressive political causes. The pursuit of aestheticism has diverted more than one leader of the 1960s. Berman has commented on the symbolic presence of the street in the radical popular music of the 1960s and early 1970s, including the music of Bob Dylan, who like Jane Jacobs began his public pilgrimage in the streets of Greenwich Village. But in the 1970s the morality tales and eschatology set to music moved in a new direction, increasingly inward, until in his film *Reynaldo and Clara* Dylan appears as writer, director, and actor, the narcissistic subject and object of a four hour event about himself (Martin 1981). The magic may be drug-induced, and the magician (as in "Mr Tambourine Man") carries his charges through a space of pure subjectivity, "where there are no fences facing." The search for an authentic symbolic repertoire remains frustrated.

The ambivalence of post-modern culture and its sponsors, the cultural new class, can move either to the political right or to the left, or indeed to a curious blend of the two, a "radical conservatism" permitting a range of outcomes. Interestingly, in the late 1970s, Bob Dylan moved away from aestheticism and at least for a period, toward traditional religious belief.[17] In this step he was joining the protest against modernity of an earlier romantic, W.B. Yeats, who asked in 1900: "How can the arts overcome the slow dying of men's hearts that we call the progress of the world, and lay their hands upon men's heart-strings again, without becoming the garment of religion as in the old times?" (Lucas 1982, p. 196). That question concerning the re-enchantment of the world remains to be answered.

Conclusion

I am well aware that this chapter has ranged far and wide–probably too far and too wide. It represents an initial attempt at an account which could be told better, and no doubt in a more focused and coherent manner. Perhaps in defence I might throw up the smokescreen that if the reader has found himself in pursuit of a moving target, well, that is the essence of modern space; if her complaint is the disarray of eclectic fragments, then she has discovered the essence of post–modern space! Yet despite all this talk about space almost nothing has been said explicitly about geography–though geographic views of space and place are implicated in most of the argument through the paper. In conclusion let me briefly draw out some of these connections more clearly.

The furious scientific empiricism of the Victorians matched their collection of domestic objects, and the desire to fill in the map coincided with the filling up of their living rooms.[18] But the meticulous description of capes, bays, and principal commodities ("everything in its place") was not purely disinterested. In its support of discovery and exploration to fill up the map, the Royal Geographical Society, for example, had one eye on imperial and commercial opportunities. Then in the 1880s and 1890s Ratzel challenged the disembodied empiricism of his contemporaries, and moved to a higher level of abstraction in his argument that geography must become a "science of distance" (Kern 1983), seeking "The laws of the spatial growth of states" (the title of an 1896 paper). His argument on the necessary relations between national growth and territorial form was imbued by the same evolutionary urgency that pervaded the spatial theory of modern architects, most overtly Louis Sullivan and his student Frank Lloyd Wright; Sullivan it was who coined the famous modernist slogan, "form follows function", an organismic slogan later to be appropriated by Nazi geopolitics. The organismic theme has been an important one in modern spatial analysis, not only indirectly through sociology and human ecology, but also directly through biology and D'Arcy Thompson's *On growth and form*, first published in 1917, which included the same abstract view of the relations between necessary function and adaptive spatial form as informed the writing of Ratzel and the architectural theorists. Thompson's analogical thinking stimulated many spatial analysts in geography including Waldo Tobler (1963), William Bunge (1964), and Peter Haggett (1965), and his concept of hexagonal symmetry may well have influenced Christaller and Losch, the founders of central place theory (Haggett 1965, p. 32, 49). This spatial thinking pursues morphological analogies without regard for the particularities of content and context. Thus in the shifting rule highways are compared to rivers; as hierarchical systems alpine glaciers and bovine livers are drawn together in a common morphological analysis; while Thompson's biological study of cell division is held to "suggest a rich field of research in the stability of geopolitical patterns over a wide time–space range" (Haggett 1969, p. 55).

The reference in all of this is clearly to Nietzsche's empty space – interestingly, Sullivan began his final book on architecture with an extract from Nietzsche - and the same modernist theory of space resonates through both modern architecture and human geography. The space of the modern architect bears a striking resemblance to the geometric orientation of the spatial analyst: for the architect "space is seen as isotropic, homogeneous in every direction....abstract, limited by boundaries or edges, and rational or logically transferable from part to whole, or whole to part" (Jencks 1981, p. 118). A shared ontology of form enabled concepts to move easily from biology to architecture and to geography – and indeed back again, for in the 1930s in preparing his urban and regional plans, Le Corbusier drew extensively upon Christaller's central place theory (Frampton 1985, p. 182).

There is no need to repeat the familiar criticism of this view of space raised within geography. It is evident enough to a non-geographer who from a survey of Ratzel's work observed "there is a distinctively abstract emphasis throughout, as if the scientific validity of his discipline depended on the reduction of geopolitical phenomena to purely spatial terms" (Kern 1983, p. 226). The suppression of local context and culture, and the imposition of uniformity as a means to universality, repeated the practices of modern architects. A similar reaction has followed in human geography with a critique derived (as in architecture and literary theory) from the philosophies of meaning; a critique whose key words would surely include the post-modern lexicon of contextuality, diversity, meaning, experience, the everyday, culture, human agency, and of course place.[19] In methodological terms the key word is interpretation. But here too there is a struggle for place, the familiar post-modern struggle of representation.[20] In discounting both a formless empiricism and formal abstraction, the task of geographical description is by no means an easy one, a task to convey the inside events and the outside causes of a place, its visible facts and less visible meanings, and all of these as they are in motion from multiple perspectives (Lewis 1985, Gregory 1986).

At present this endeavour hangs in the balance. It contains the possibility of a revived and creative human geography built around a newly informed synthesis of people and place. But it contains also the possibility of the same radical subjectivity and individualism that has sometimes diverted the post-modern project in other fields. In the work of human geography's most brilliant experimentalist, Gunnar Olsson, a surrealist attack on categories, the celebration of ambiguity ends up as an often closed hermetic discourse. A radical attempt at representation becomes self-referential, creating a private world, and taken to its conclusion leads to the anarchism of the smallest community in the smallest space. As Olsson (1979) puts it, quoting Appollinaire: "There are only two of us in the cell: I and my mind."

Acknowledgements

This chapter is a revised and expanded version of a paper presented to a conference organized by the Center for Humanistic Social Science at the University of Waterloo, to the Historical Geography seminar at Cambridge University, and in a fuller form to the Department of Geography at

Syracuse University, all in 1986. It has benefitted from comments received at all three places. While the theoretical literature on post-modernism has expanded mightily in the period since this paper was first presented – see for example the special double issue of *Theory, Culture and Society* (June 1988) – I have chosen to let the argument stand in its original form. So too the growing penetration of post-modernism as a serious and successful design form has continued apace – for commercial building see, for example, Chao & Abramson (1987) on the Kohn Pedersen Fox partnership, and for residential building, the review of American neo-vernacular landscapes in Langdon (1988).

Notes

1 Consider among the modernists in architecture, Mies van der Rohe (1926) (in Giedion 1967): "Architecture is the will of the epoch translated into space"; Giedion (1938–9) (in Giedion 1967) "It is obvious that in the second decade of this century the same spirit emerged in different forms, in different spheres, and in totally different countries"; an appeal to the spirit of the age is pervasive in Gropius, the organicism of Frank Lloyd Wright, and the writing of Le Corbusier, for example in the title of his periodical, *L'esprit nouveau*.

2 For additional complaints directed at late Victorian home decoration by Voysey and Giedion, see Kern (1983, p. 156).

3 Yeats was both a symbolist and a defender of religious values, positions which increasingly isolated him in literary and intellectual circles: see Lucas (1982, Ch.9).

4 Such adjectives as 'honest', 'sincere', and 'pure' are favorite descriptive terms of the modernists to refer to their own work, implying also of course a deliberate moral judgement of other traditions. Giedion (1967) a student of Le Corbusier and perhaps the most successful disseminator of modern design theory, has a section entitled "The demand for morality in architecture" in his book, *Space, time and architecture*.

5 The relations between these two prophets of modernism are close, and include Le Corbusier's role in reprinting the French translation of *Ornament and crime*; see Frampton (1985, p. 95).

6 This continues to be true of contemporary modernists like Arthur Erickson, who justified a recent design in terms of its view from an airplane. See the discussion in Ley (1987).

7 In 1913 in the midst of a period of factory design, Gropius (later to be director of the Bauhaus), found time to design a diesel locomotive and railway sleeping car; Frampton (1985, p. 114).

8 The theme of establishing a new social order as a central tenet of modernism is developed further in Ley (1987).

9 See Giedion (1967) for example pp. 317, 433–4; Zorbaugh (1929), pp. 271–5.

10 As stated in a television interview in the British television series, *Architecture at the crossroads*, BBC 2, March 1986.

11 Thus for Habermas and Berman the post-modern is seen as a moment of the modern and incorporated within it. This is a single instance of a widely varying and confused use of the terms modern and post-modern.

12 The townhouse is in the Fairview Slopes district, Vancouver. See also the critical assessment of the new Broadway Office Centre, West London, "an

extraordinary mixture of oriel windows, gables, Art Nouveau, Art Deco, red-brick towers, 'chateaux' roofs, Tudorbethan 'farm' buildings and conservatory structures" (Gardiner 1985).

13 Symptomatic is the controversy around the Terry Fox memorial in Vancouver. In its adaptation of a triumphal arch, the memorial draws upon a populist post-modernism to commemorate the young one-legged runner who died while running across Canada raising money for cancer research. The memorial was roundly condemned as cheap and vulgar, disrespectful of a heroic feat. Vancouver City Council passed a motion to demolish it. It raises the question of whether heroic architecture is possible in a sceptical, consumer culture.

14 This restatement of Adorno's thesis appears in Martin (1981); for a current example see Ley (1980). For a historic precedent consider the frustrated exclamation of C.R. Ashbee, one of William Morris' contemporaries in the arts and crafts movement: "We have made of a great social movement, a narrow and tiresome little aristocracy working with great skill for the very rich." Cited in Harris (1984). The affinities between post-modernism and the late 19th-century symbolist/aesthetic movement could be usefully explored. See for example the parallels between Sitte, Ruskin, and Jane Jacobs suggested in the following section of the chapter.

15 Peter Fuller (1984) identifies Ruskin as, "the true prophet of the 'post-modern' and 'post-industrial' era". I am indebted to a forthcoming essay "Iconography and landscape" by Stephen Daniels and Denis Cosgrove for this reference.

16 Berman (1982, p. 13) writes that "To be modern...is to be both revolutionary and conservative"; compare Daniel Bell's self-description as "a socialist in economics, a liberal in politics, and a conservative in culture." See also Fuller (1984). There is clearly need for further thought on this blurring of conventional categories.

17 For example in his ironic treatment of narcissism (and its manipulators) in "License to kill" on the album, *Infidels* (1983). Dylan has increasingly become a master of irony, an important post-modern genre.

18 Here as elsewhere, Ruskin was a characteristic Victorian: compare "his great eclectic collections or 'cabinets' of materials – mineral, floral and artifactual specimens, so typical of the Victorian intellectual sensibility." Daniels & Cosgrove forthcoming.

19 Though different authors vary considerably in their relations to philosophical idealism, this lexicon runs pervasively through a broadly defined cultural–humanistic geography. See, for example, Tuan (1977), Ley & Samuels (1978), Cosgrove (1984).

20 In architecture, see Jencks (1983); in literary theory, Arac (1986); in anthropology, Marcus & Fischer (1986) (especially Ch. 1 "A crisis of representation in the human sciences").

References

Agena, K. 1983. The return of enchantment. *New York Times Magazine*, 27 November.

Appleyard, D. 1981. *Livable streets.* Berkeley: University of California Press.

Arac, J. 1986. Introduction. In *Post-modernism and politics*, ed. J. Arac, pp. ix–xliii. Minneapolis: University of Minnesota Press.

Barnard, P. 1984. *A short history of English literature.* Oxford: Blackwell.

Bell, D. 1976. *The cultural contradictions of capitalism.* New York: Basic Books.

Bell, D. 1980. *The winding passage*. New York: Basic Books.

Berman, M. 1982. *All that is solid melts into air: the experience of modernity*. New York: Simon & Schuster.

Bernstein, R., ed., 1985a. *Habermas and modernity*. Cambridge: Polity Press.

Bernstein, R. 1985b. Introduction. In *Habermas and modernity*, R. Bernstein (ed.), (pp. 1–32. Cambridge: Polity Press.

Boyer, M. 1983. *Dreaming the rational city*. Cambridge: MIT Press.

Branzi, A. 1984. *The hot house: Italian new wave design*. Cambridge: MIT Press.

Brolin, B. 1976. *The failure of modern architecture*. London: Studio Vista.

Bunge, W. 1964. Patterns of location. *Michigan Inter-University Community of Mathematical Geographers* No. 3.

Chao, S. & T. Abramson, (eds) 1987. *Kohn Pedersen Fox*. New York: Rizzoli.

Collins, G. & C. Collins 1965. *Camillo Sitte and the birth of modern city planning*. New York: Random House.

Cosgrove, D. 1984. *Social formation and symbolic landscape*. London: Croom Helm.

De Lacy, J. 1983. Cultivating culture in Paris. *New York Times Magazine*, 22 May.

Edgar, D. 1986. It wasn't so naff in the 60s after all. *The Guardian*, 7 July, 21.

Foster, H. 1983. Postmodernism: a preface. In *The anti-aesthetic: essays on postmodern culture*, H. Foster (ed.), pp. ix–xvi. Port Townsend, Wash.: Bay Press.

Frampton, K. 1983. Towards a critical regionalism. In *The anti-aesthetic: essays on postmodern culture*, H. Foster (ed.), pp. 16-30. Port Townsend, WA: Bay Press.

Frampton, K. 1985. *Modern architecture: a critical history*. London: Thames & Hudson.

Fuller, P. 1984. John Ruskin: a radical conservative. In *Images of God: the consolation of lost illusions*, P. Fuller (ed.), pp. 277–83. London: Jonathan Cape.

Gardiner, S. 1985. Mad world. *The Observer*, 24 March, 23.

Giedion, S. 1967. *Space, time and architecture*, 5th edn. Cambridge: Harvard University Press.

Gold, J. 1985. From 'Metropolis' to 'The City': film visions of the future city, 1919–1939. In *Geography, the media and popular culture* J. Burgess & J. Gold (eds), pp. 123–143. London: Croom Helm.

Gregory, D. 1986. Areal differentiation and post-modern human geography. Unpublished paper, Department of Geography, Cambridge University.

Habermas, J. 1983. Modernity: an incomplete project. In *The anti-aesthetic: essays on postmodern culture*, H. Foster (ed.), pp. 3–15. Port Townsend, WA.: Bay Press.

Haggett, P. 1965. *Locational analysis in human geography*. London: Arnold.

Haggett, P. 1969. *Network analysis in geography*. London: Arnold.

Harris, W. 1984. An anatomy of aestheticism. In *Victorian literature and society*, J. Kincaid & A. Kuhn (eds), pp. 331–47. Columbus: Ohio State University Press.

Iglauer, E. 1981. *Seven stones: a portrait of Arthur Erickson*. Madeira Park, BC: Harbour Publishing.

Jacobs, J. 1964. *The death and life of great American cities*. Harmondsworth: Penguin.

Jameson, F. 1983. Postmodernism and consumer society. In *The anti-aesthetic: essays on postmodern culture*, H. Foster (ed.), pp. 111–25. Port Townsend, WA: Bay Press.

Janik, A. & S. Toulmin, 1973. *Wittgenstein's Vienna*. New York: Simon & Schuster.

Jay, M. 1985. Habermas and modernism. In *Habermas and modernity*, R. Bernstein (ed.), pp. 125–39. Cambridge: Polity Press.

Jencks, C. 1980. *Late-modern architecture and other essays*. New York: Rizzoli.

Jencks, C. 1981. *The language of post-modern architecture*. New York: Rizzoli.
Jencks, C. 1983. The perennial architectural debate. In *Abstract representation*, C. Jencks (ed.), pp. 4–22. London: Architectural Design Profile, Academy Editions.
Jencks, C. 1985a. Ornament and symbolism in post-modern art. Lecture presented to the Emily Carr College of Art and Design, Vancouver.
Jencks, C. 1985b. Star struck house. *House and Garden*, April, 112 et seq.
Jencks, C. & W. Chaitkin, 1982. *Architecture today*. New York: Abrams.
Kern, S. 1983. *The culture of time and space 1880–1918*. Cambridge: Harvard University Press.
Knox, P. 1987. The social production of the built environment: architects, architecture and the post-modern city. *Progress in Human Geography* **11**, 354–77.
LaCapra, D. 1983. Rethinking intellectual history and reading texts. In *Rethinking intellectual history: texts, contexts, language*, pp. 23–71. Ithaca, NY: Cornell University Press.
Langdon, P. 1988. A good place to live. *The Atlantic* **261** (3): 39 et seq.
Lasch, C. 1978. *The culture of narcissism*. New York: Norton.
Lears, T. J. J. 1983. From salvation to self-realization: advertising and the therapeutic roots of the consumer culture, 1880-1930. In *The culture of consumption*, R. Fox & T. J. J. Lears (eds), pp. 3–38. New York: Pantheon.
Le Corbusier, 1927. *Towards a new architecture*. London: John Rodker.
Leitner, B. 1976. *The architecture of Ludwig Wittgenstein*. New York: New York University Press.
Lewis, P. 1985. Beyond description. *Annals, Association of American Geographers* **75**, 465–78.
Ley, D. 1980. Liberal ideology and the post industrial city. *Annals, Association of American Geographers* **70**, 238–58.
Ley, D. 1985a. Cultural–humanistic geography. *Progress in Human Geography* **9**, 415–23.
Ley, D. 1985b. *Gentrification in Canadian inner cities: patterns, analysis, impacts and policy*. Ottawa: Canada Housing and Mortgage Corporation.
Ley, D. 1987. Styles of the times: liberal and neo-conservative landscapes in inner Vancouver, 1968–1986. *Journal of Historical Geography* **14, 40–56**.
Ley, D. & M. Samuels, 1978. *Humanistic Geography*. Chicago: Maaroufa.
Lorimer, J. 1983. Citizens and the corporate development of the contemporary Canadian city. *Urban History Review* **12**, pp. 3–9
Lucas, J. 1982. *Romantic to modern literature: essays and ideas of culture*. Brighton, Sussex: Harvester Press.
Lyotard, J-F 1984. *The postmodern condition*. Manchester: Manchester University Press.
Marcus, G. and M. Fischer, 1986. *Anthropology as cultural critique*. Chicago: University of Chicago Press.
Martin, B. 1981. *A sociology of contemporary cultural change*. Oxford: Blackwell.
McAfee, A. 1983. *The renewed inner city neighbourhood – is one out of three sufficient?* Vancouver: City of Vancouver Planning Department.
Miller, J. Hillis 1963. *The disappearance of God: five 19th century writers*. Cambridge: Harvard University Press.
Olsson, G. 1979. Social science and human action or on hitting your head against the ceiling of language. In *Philosophy in geography*, S. Gale & G. Olsson (eds), pp. 287–307. Dordrecht: Reidel Publishing Company.
Park, R. 1925. Magic, mentality and city life. In *The city*, R. Park, E. Burgess & R. McKenzie (eds), pp. 123–41. Chicago: University of Chicago Press.

Poster, M. 1975. *Existential marxism in postwar France*. Princeton: Princeton University Press.

Relph, E. 1976. *Place and placelessness*. London: Pion.

Schorske, C. 1981. *Fin-de-siècle Vienna: politics and culture*. Cambridge: Cambridge University Press.

Scott Brown, D. 1980. Architectural taste in a pluralistic society. *Harvard Architecture Review* 1, pp. 41–51.

Time Magazine, 1985. New gilded age grandeur. 2 September, pp. 34–5.

Tobler, W. 1963. D'Arcy Thompson and the analysis of growth and form. *Papers of the Michigan Academy of Science, Arts and Letters* **48**, pp. 385–90.

Tuan, Y-F 1977. *Space and place: the perspective of experience*. Minneapolis: University of Minnesota Press.

Van Nus, W. 1979. Towards the city efficient: the theory and practice of zoning. In *The usable urban past: planning and politics in the modern Canadian city*, A. Artibise & G. Stelter (eds), pp. 226–46. Toronto: Macmillan.

Von Wright, G. 1984. Biographical sketch. In N. Malcolm, *Ludwig Wittgenstein. a memoir*. London: Oxford University Press.

Venturi, R., Scott Brown, D., & Izenour, S. 1972. *Learning from Las Vegas*. Cambridge: MIT Press.

Watkin, D. 1977. *Morality and architecture*. Oxford: Clarendon.

Wernick, A. 1984. Sign and commodity: aspects of the cultural dynamic of advanced capitalism. *Canadian Journal of Political and Social Theory* **8**, pp. 7–34.

Wirth, L. 1925. A bibliography of the urban community. In *The city*, R. Park, E. Burgess & R. McKenzie (eds), pp. 161–228. Chicago: University of Chicago Press.

Wirth, L. 1964. *On cities and social life*, A. Reiss (ed.). Chicago: University of Chicago Press.

Wright,, F. Lloyd 1975. *In the cause of architecture*. New York: Architectural Record Books.

Zorbaugh, H. 1929. *The gold coast and the slum*, Chicago: University of Chicago Press.

5

Home and class among an American landed elite

PETER J. HUGILL

How America came to be "home" for her population is a major, largely unexplored theme in American history. With the Revolutionary War successfully concluded it became clear that Americans were no longer British, but then what were they? Much thought was devoted to the problem as a democratic republic emerged, at least in the northeastern states. The Civil War, however, shattered these republican interpretations of home and resulted in an elite reversion to Anglicization which lasted until the end of World War I. Only after World War I did a powerful mass culture emerge that was genuinely American in nature, and that was predicated upon the technical changes of national radio broadcasting, cinema, and the automobile as much as on more traditional cultural values.

This chapter concerns the first two phases of America's becoming "home" to the elite of the northeast: the republican phase before the Civil War, and the period of re-Anglicization between then and World War I.

In the period immediately following the Civil War the regional elite of the northeastern American states became a national elite. What had been in part a society in which wealth was derived from agricultural production became a society in which industry became the source of wealth. The growth of industry created a tremendous demand for labor which could not be met internally. Large numbers of immigrants began to pour into America's cities from first western, then southern and eastern Europe. What had been a largely homogenous, white Anglo-Saxon Protestant, English-speaking region of America was increasingly transformed into a polyglot culture. This was particularly problematic for the old-stock elite. Industrial growth was providing hitherto unheard-of wealth, but the roots of old-stock culture were being progressively undermined in the industrial cities. The old-stock elite solved the problem in two ways: they returned both culturally and physically to their English roots, making summer pilgrimages to England (Lockwood 1981, p. 288) and de-emphasizing the democratic "Americanness" they had sought to create during the republican period of American history from the Revolutionary War through the Civil War; and they fashioned a distinctively American version of the English landed elite, one which was less rooted to the countryside and did not need large incomes from the land. Even the roots of this were English, as English industrialists assiduously sought to invest their profits in land for status reasons, even when the return on their capital was poor (Wiener 1981, p. 12).

These forces are revealed at work in a case study of a small village in Upstate New York, Cazenovia. Before the Civil War, Cazenovia was home to a group of families that personified "New England Extended." Settlement began immediately the Indian problem was settled in Upstate New York. The family that most came to define Cazenovia as "home" was central to the village's settlement in the 1790s. The Lincklaen-Ledyard family epitomized the established values of the elite of republican America. Internationalist: Lincklaen, who married Helen Ledyard in 1797, was the agent for four Dutch banking houses that sought profit from land sales in an America hungry for development capital. Patriotic: Helen Ledyard's family lost five officers in the Revolutionary War. Tolerant of other cultures: Lincklaen, a well-traveled officer in the Dutch Navy before settling in America, was as fluent in French and English as in his native Dutch. Sophisticated: Lincklaen wintered in New York, yet built houses, churches, and courthouses on the New York frontier of the 1790s that were the equal of any in America and many in Europe. Democratic: the Lincklaen-Ledyards sold land, never leased it, and were generous about repayments from their farmer customers when times were hard. Self-effacing: for all their local prestige and power they routinely appear in the census as "farmers" when other local families described themselves as "gentlemen" or "capitalists", and they seem to have usually resisted using their considerable influence in Albany to manipulate the affairs of the state in their favor.

Before the Civil War the Lincklaen-Ledyards lived comfortably in Cazenovia on a combination of profits from the sale of land and judicious investments in land and small-scale manufacturing in Cazenovia and elsewhere. Lincklaen had bought out the four Dutch banks in 1816 and the debt to them was paid off by 1841 by his heir and brother-in-law, Jonathan D. Ledyard. After Lincklaen's death in 1822 Jonathan D. Ledyard worked hard at improving the village, making it even more a long-term home for his family. It was, however, his eldest son, who later reversed his name to Ledyard Lincklaen to please his aunt, John Lincklaen's widow, who first codified in writing what it was that made Cazenovia home to he and his kin. Ledyard Lincklaen made much of Cazenovia's "snugness", in particular the "snugness" of its landscape and the integration of the house into that landscape. He was the first Lincklaen-Ledyard to travel to England in the republican period, in 1852, and he praised English rural tastes. In a review of Susan Fenimore Cooper's *Rural Hours* he compared English and American tastes while praising Cooper for "the first American book . . . devoted to rural life and the objects of the country" (Lincklaen 1855, p. 32). Lincklaen's analysis was very much concerned with what makes a home homely. Dismissing England's cities as "not unlike our own" (Lincklaen 1855, p. 33) he concentrated on the appearance of the rural landscape and the rural dwelling. To Lincklaen (1855 p. 33),

so far as appearances are concerned, the English cottage has an air of snugness and shelter woefully wanting in the Yankee house

. . . And then our better houses, close to the highway, erect, stiff, sharp-cornered, full of windows as a lantern . . . they are sufficiently comfortable, but they too often suggest a chill and shiver to the passer-by on a winter day.

Much of this Lincklaen attributed to the age of England's landscape, contrasting an England well-settled even in Roman times with an America that traces her antiquity only to "an old house built of Dutch bricks . . . , a mansion temporarily occupied by some general officer in the Revolutionary war . . . the Indian flint arrowheads ploughed up in our fields" (Lincklaen, 1855, p. 34). He praised Cooper for her identification of distinctively American features in flora and fauna, and suggested many himself, though almost always in comparison with English examples. Thus, to give but two examples, "we surpass England in the number of our swallows" (Lincklaen 1855, p. 37) and "our trout and salmon are as fine as those of Scotland" (p. 38). He concluded by identifying the well-educated Englishwoman with her naturalist knowledge as the focal point of these sensibilities (Lincklaen 1855, p. 40):

If there is any portion of the English people possessed of a real superiority, and suitable for a model to others, it may be confessed to be the best and most intelligent class of Englishwomen not involved in gay and fashionable life. Among these, there is no more marked and healthful peculiarity than their intimate knowledge of and interest in the animated races and plants and trees which surround their beautiful homes, whether domesticated favorites of the farm and garden, or wild denizens of forest and field, and the staunchest native American, or the most sensitive patriot, cannot object to our injunction on the ladies of this country, that they should imitate such examples. . . .

I quote Ledyard Lincklaen at length since he epitomized the forces at work in republican America that, attempting to define America as home, laid the groundwork for the increasing Anglophilia of the American elite after the Civil War while himself avoiding such sentiments directly. Confronted with the new immigrants the elite retreated to the countryside, which it viewed through increasingly English eyes (Hugill 1986). We may identify four themes in Lincklaen's commentary and follow those through the post-Civil War period as the Anglophilic American landed elite defined itself: a concern with the picturesque in settlement, house, and landscape; a search for antiquity; a concern for the natural world; and the notion that women are the guardians of these values.

In the last category Lincklaen clearly included Susan Fenimore Cooper. More than any other work except Thoreau's *Walden*, Cooper's *Rural Hours* defined landscape sensibilities in the northeast before the Civil War. First published in 1850, *Rural Hours* sold well from the start, both in America and in England. It was reprinted several times, being shortened by some 200 pages for the final, revised edition of 1887. Cooper showed a strong concern with the role of water in defining the picturesque: "it is an essential part of prospects widely different in character" (Cooper 1968, p. 75). Of Cooperstown's location on Lake Otsego she commented that (Cooper 1968, pp. 75–6):

> the lake, with its clear, placid waters, lies gracefully beneath the mountains, flowing here into a quiet little bay, there skirting a wooded point, filling its ample basin, without encroaching on its banks by a rood of marsh or bog.
>
> And then the village, with its buildings and gardens covering the level bank to the southward, is charmingly placed, the waters spreading before it, a ridge of hills rising on either side, this almost wholly wooded, that partly tilled, while beyond lies a background, varied by nearer and farther heights.

Cooper amplified this concern with the picturesque in a long description of a farmhouse (1968, p. 104):

> From the window of the room in which we were sitting, we looked over the whole of Mr. B's farm; the wheat-field, corn-field, orchard, potato-patch, and buckwheat field. The farmer himself with his wagons and horses, a boy and a man, were busy in a hay-field, just below the house; several cows were feeding in the meadow, and about fifty sheep were nibbling on the hillside. A piece of woodland was pointed out on the height above, which supplied the house with fuel.

Cooper particularly insisted on the role of trees, and thus of nature, in the creation of the picturesque landscape (1968, p. 153).

> a fine tree near a house is a much greater embellishment than the thickest coat of paint that could be put on its walls, or a whole row of wooden columns to adorn its front; nay, a large shady tree in a door-yard is much more desirable than the most expensive mahogany and velvet sofa in the parlor.

After complaining that the first tendency of American settlers was to cut down trees Cooper linked "preservation of fine trees" to "the civilization of a country: they have their importance in an intellectual and in a moral sense" (1968, p. 153).

Apart from her concerns for the picturesque in settlement, house, and landscape, and for the natural world, Cooper was also concerned with antiquity (1968, p. 95):

> The fields which border this quiet bit of road are among the oldest in our neighborhood . . . one might readily believe these lands had been under cultivation for ages . . . a stranger moving along the highway looks in vain for any striking signs of a new country. . . . Probably there is no part of the earth, within the limits of a temperate climate, which has taken the aspect of an old country so soon as our native land.

Thus in both Cooper's work and Lincklaen's response the same themes dominate. Through these themes they, and a host of less literate individuals, strove to domesticate the northeast of America, in particular that region west of the Berkshires that their parents' generation had pioneered. Before the Civil War these themes were couched in generally American contexts, but after the war a new generation reconceived them in anti-immigrant, anti-urban, and Anglophilic contexts.

Ledyard Lincklaen was the eldest of a large family. Upon his death in 1864 at age 44 it was left to his youngest brother, Lambertus Wolters Ledyard (LWL), to be his family's voice in the day-to-day running of Cazenovia. In many ways it was LWL who transformed Cazenovia from a home for the Lincklaen-Ledyard family into a home for some of the newly rich elite of the northeast. It was certainly LWL who pounded the anti-immigrant, anti-urban, and Anglophilic drums.

In purely local terms the immediate cause of the beating of these drums was the coming to the village of two railroad lines, one from Canastota in 1870, the other from Syracuse in 1872. The immediate impact of these was to destroy the old manufacturing economy. By 1875 the taker of the state census could record only 25 persons employed in industry, with the comment that (State Census of NY, County of Madison, Section V, 1875):

> Most all manufacturing in this town has been done away with and the old buildings are used by one or two persons with each building manufacturing one thing and another.

The railroads also seemed to provide a way out of this crisis. Two major picnic grounds were opened on the shores of Cazenovia Lake, and steamboat service between them and the village proliferated after 1873. As LWL retrospectively put it (Cazenovia *Republican* – henceforth CR – March 3 1887, "The Sunday Question"):

> Cazenovia would seem to have experimented in these ways so far as keen business judgment would suggest. Almost everything has been worked for money. Temperance for drinks, camp-meeting religion for money at the gate, Sunday-school assemblys [*sic*] for the good of real estate, the beauty of the lake for noise and confusion, and no end of schemes more or less unsuccessful.

All this LWL clearly related to problems with the city and the immigrant (CR, March 3 1887):

> Now, will we help Cazenovia by introducing the unlimited beer and tobacco Sunday that the Germans have planted to perfection in the Bowery parts of New York, and still more firmly in Chicago?

The end results LWL saw equally clearly (CR, January 26 1888, "Village Expenditures"):

Few of the great cities of the United States are especially handsome, hardly any are well governed or encourage municipal integrity, they are generally more or less unhealthy and a large part of the most successful men in them have plans for eventually getting homes in some attractive country place . . . where, amid pleasant surroundings and in safety, their children can grow brown in the sun and sturdy with open air activity.

Fine words and knowing ones indeed from a man whose aunt's grandson, niece's husband, and neighbor, Charles Stebbins Fairchild, was the State Attorney General who broke the Tweed Ring.

LWL capitalized on this urban fear, as well as on Cazenovia's natural beauties which he conscientiously sought to improve by private and public expenditure. Before the real impact of the railroads was felt the *New York Times* could refer to Cazenovia as a cool and pleasant retreat (reprinted CR, 7–31–1873),

its first charm still untouched by the follies of fashion. A few stray visitors from year to year have alone discovered the beauties and attractions of the place as a summer resort. . . . [It] would soon become a noted and famous resort if the people here wanted it – as they do not – overrun with strangers every summer.

The *New York Times* article was by no means unique. Recognition of Cazenovia's attractions was clear in numerous articles published elsewhere and frequently reprinted in the *Republican*. All stressed the beauty of the Cazenovia landscape, its quiet, and its social advantages. The New York *Evangelist* (CR, 11–2–1867) and the *Oxford* [NY] *Times* (CR, 11–6–1867) in 1867; the *New York Tribune* (CR, 8–10–1870) and the Trenton, NJ, *State Gazette* (CR, 8–24–1870) in 1870; the Roundout, NY, *Freeman* (CR, 10–24–1872) in 1872; the *New York Times* (CR, 7–31–1873) and the *Syracuse Standard* (CR, 8–7–1873) in 1873; the *Standard* again in 1875 (CR, 4–15–1875); the New York *Mail* in 1876 (CR, 9–21–1876); and the New York *Daily Graphic* on April 9, 1878, with sketches. This outside praise ceases at this point, with the exception of an article entitled "A Modern Utopia" from the *Syracuse Journal* of 1885 (CR, 8–30–1885), probably because of the success of the day trippers reconstructing Cazenovia to the point where it was no longer quiet and beautiful. It resumes properly on August 6, 1889, the year the last steamboat left the lake, with an article in the *Syracuse Standard*, followed in 1890 by an article entitled "An Acadian Village" in the New York *Journal* (CR, 6–12–1890) and one on "Beautiful Owagwena" in the *Syracuse Journal* (CR, 7–17–1890). The New York *Mercury* (CR, 8–8–1895) and the *Delaware Republican* of Delhi, NY (CR, 8–8–1895) had articles in 1895; as did the Syracuse *Post Standard* of May 25, 1899.

LWL had much in person to do with this recovery. He entered village politics with a vengeance, being elected president in 1886, 1887, and

1889, passed a stringent ordinance against the blowing of steamboat whistles, and committed Cazenovia to a course of public expenditure on sewers, water supply, roads, and aesthetic improvements designed to make the village more attractive than ever to the elite. The latter objectives were obtained through a veritable barrage of articles in the *Cazenovia Republican* to educate the public to the advantages of such a course (CR 3–13–1884, 4–10–1884, 4–24–1884, 6–19–1884, 7–30–1884, 7–30–1885, 2–3–1887, 2–24–1887, 3–3–1887, 3–10–1887, 3–31–1887, 1–18–1888, 1–26–1888, 2–9–1888, 3–1–1888).

By 1888 it was clear that Cazenovia was to be a home to the elite. LWL summed up the characteristics of such folk succinctly (CR, 1–26–1888, "Village Expenditures").

> The families who are inclined to found homes in such communities as Cazenovia are among the most conservative and prudent of American Society, and the place may be regarded as fortunate that possesses the elements that are attractive to men of wide experience, extensive travel, and liberal means. The characteristics that enhance the pleasure and prosperity of every person living in Cazenovia, are the ones that will add to our population the most desirable class of strangers, and also fit our village more perfectly for the requirements of students and persons retiring from farms and business.

Attracting such folk required considerable expenditure (CR, 1–26–1888).

> Less than three years have added to Cazenovia, the forest park, the new pier, the public library, the "Owahgena" [club], the handsomely refitted Lincklaen House [hotel], the "Casanova" [opera house], the Fairchild mechanical building, stone roads, new street coping, beautiful drives on the lake with many attractive residences in the village, valuable additions to the seminary, and many other improvements, yet people often charge that this place is sluggish and perchance is so regarded because all has been smoothly and quietly done.

LWL was solely responsible for the construction of the "Owaghena" Club (CR, 5–7–1885), which catered only to the most elite summer residents, as selected by LWL. The rebuilding of the "Casanova" opera house he shared with his brother, George Strawbridge Ledyard (CR, 2–22–1883). The street coping (CR, 10–22–1885), the new pier (CR, 11–23–1893), and the stone crusher to produce the stone for the stone roads (CR, 12–3–1885, 3–25–1886, 7–15–1886, 1–10–1889) were elements initially provided by LWL to be paid for by the village only as and when the taxpayers approved, which they invariably did. He played a leading role in the formation of the library association (CR, 1–23–1873), and built many of the attractive new residences mentioned (CR, 11–22–1877, 5–1–1877, 12–17–1885, 6–2–1887). Later evidence shows he must have been involved in some of the lakeside drives (CR, 7–13–1893, 10–12–1893), so that only in the provision of the forest park, the seminary improvements, and the Fairchild Mechanical building is his involvement in doubt, and the Fairchilds, in any case, were a directly related family.

He also was responsible for less important elements of the landscape. Water troughs and lamps (CR, 6–3–1875), a rustic fountain (CR, 4–26–1883), and, with R. J. Hubbard, a new boathouse at the Pier (CR, 7–22–1886). More important, perhaps, was his institution of the Lake Fete in 1877 (CR, 8–30–1877). Fourteen such Fetes were held between 1877 and 1889, while LWL was regaining and consolidating his control.

During his terms as village president LWL also used his office to persuade others to improve their landscape. In 1886 he persuaded the then Elmira, Cortland, and Northern Railroad, heirs to the Cazenovia and Canastota, to create a landscaped circular driveway at their station for carriages to turn around (CR, 5–20–1886).

This was round one, as it were. Round two called for more expenditures than even the wealthiest of Cazenovia families could afford. LWL summarized the situation before beginning his third term as president (CR, 1–26–1888).

> Conscientious men have guided the affairs of the village from its settlement, and a vast amount of good, tasteful, well judged work is being done, as a glance through our streets and public places will show. All that is needed for the future is to keep alive to important needs, and not hesitate too long in consenting to such public expenditure as may in the end prove actual economy in the preventing of fires, illness, and consequent loss of reputation.

These were forthcoming, both from the public purse and from private purses of much greater depth than those of the Lincklaen–Ledyards. In 1891 Henry Burden II first came to Cazenovia, renting "Corner Cottage" from LWL (CR, 4–30–1891). Burden's father, a Scottish immigrant, made a great fortune in the Civil War as proprietor of the Troy (NY) Iron and Nail Works, the chief supplier of horseshoes to the Union Army. More than that, he devised a standard sizing system for horseshoes, so that blacksmiths could fit individual horses almost "off-the-peg."

Burden was not always well-accepted by the old elite of the village, perhaps because of his assertiveness (Grills 1977, p. 66), perhaps because he was not of an old New England family. Nevertheless he vigorously "improved" the village and was socially rewarded for it. In 1892 he contributed generously to LWL's scheme to improve local roads with a local stone crusher (CR, 10–6–1892). In 1894 Mrs. Burden paid for "2,500 feet of fine gravel walk, three and half feet wide" on the west side of Lake Road (CR, 5–26–1894). That same year Henry Burden was elected to the House Committee of the Cazenovia Club (CR, 7–26–1894). In 1895 the Burdens planted geraniums and vines in boxes on Albany Street (CR, 6–6–1895), built "an expensive boathouse," (CR, 7–11–1895) and put a great deal of effort into the road near the head of the lake, for which district Henry was pathmaster (CR, 10–17–1895): he was also re-elected to the Cazenovia Club's House Committee (CR, 8–1–1895). In February of

1896 Henry stepped outside his role as summer visitor and came up from New York the early part of the week to be present at the town meeting (CR, 2–13–1896).

The next year Burden put up $5,000 for the construction of a new opera house (CR, 1–1–1897) to help replace the Lincklaen-Ledyard family's "Casanova," destroyed by arson in 1895 (CR, 10–31–1895). Burden received $1,000 in stock in the new opera house for his generosity. The Lincklaen-Ledyard family came in belatedly, with Benjamin Brewster's contribution (CR, 9–23–1897). That same year Henry Burden found himself on the executive committee of Charles S. Fairchild's new golf club alongside Mr Wendell and a Mr Leech. Both Fairchild and Wendell were Lincklaen-Ledyards. The paper commented (CR, 9–19–1897):

> Golf in Cazenovia this season is quite the rage, and the links on the Honorable Charles S. Fairchild's estate are crowded with players and aspirants. Two years ago saw but a handful of players, Mr. J.F. Leech (captain of the crack Washington, D.C., club) being the prominent pioneer.

Burden also "donated $90,000 as a matching grant to build Centennial Hall at Cazenovia Seminary," and purchased and renovated several important old buildings, including the "Lincklaen House" Hotel (Grills 1977, p. 66). Perhaps his major contribution to the appearance of the village was to buy up the Union Electric Company in 1902 and, as president of the new Cazenovia Electric Company, replace all the overhead with underground wiring (Niagara Mohawk Archives, 6–10–1905). The expense of such underground installation was generally considered ruinously expensive by contemporaries (*Electrical World*, 7–5–1913).

Burden was certainly not alone, although he was the most aggressive and possibly the wealthiest of Cazenovia's elite by the end of the century. Other wealthy folk of New England descent, many of them related by marriage to the Lincklaen-Ledyards, bought land on the lake or close to it, built imposing mansions, and used Cazenovia as a summer or retirement home. None of these folk made their living in Cazenovia. Two of many possible examples illustrate this clearly. Benjamin Brewster built "Scrooby" on the southern edge of the lake between 1888 and 1890 (CR, 5–8–1890) at a cost reputed to have been 1 million dollars (CR, 6–7–1894). Brewster was related to the Lincklaen-Ledyards by marriage, and was one of John D. Rockefeller's earlier partners in Standard Oil (Hawke 1980, p. 177). Robert Benson Davis was not related to the Lincklaen-Ledyards: in 1905 he built "Hillcrest" overlooking Cazenovia Lake from the top of the east slope. His wealth came from the Davis Balking Powder Company, of which he was the founder (Grills, 1977, p. 70).

In fact, by 1899 a reporter for the Syracuse *Post Standard* could write, without too much hyperbole (5–22–1899):

> Cazenovia can properly be called the summer home of millionaires. The statement is made, and there appears nobody to dispute it, that

when the season is at its height and all those who generally attend are present, the congregation of St Peters Episcopal church represents $100,000,000, yet the congregation is not a large one.

The consolidation of the elite into the late 19th century summer colony was retrospectively summed up in 1936 (CR, 6–18–1936):

Gentlemen in cutaway coats, striped trousers and stiff hats, have leisure to discuss politics, religion or anything else as they meet on the sidewalk; the word, "hello" is rarely heard, even from young people addressing their elders; it is "good morning" or "good evening" spoken distinctively; there are lifted hats, and salutes, sometimes with gold headed canes. School teachers and professors, as well as ministers and rectors and priests, are treated with great deference, even on questions of national policy. President Cleveland's controversy with the Senate has brought out the phase "innocuous desuetude", the populism and anarchy of the west are dreaded and feared, "fiat money" is warmly debated, the railroads, Standard Oil and all monopolies are regarded as public enemies, – unless one happens to be a stockholder.

This emergent elite was a selective one and very English in its social characteristics. New money was clearly frowned upon, which explained the lack of social acceptance of such folk as Henry Burden. Even minor misbehavior was enough excuse to cut undesirables. In 1888 the *Cazenovia Republican* reported, in marvelously Victorian style, the divorce from Jacob Vanderbilt, Jr, of a Cazenovia woman of Irish ancestry who used to wait tables at the Lincklaen House Hotel with the comment that (CR, 5–24–1888)

Mr. Vanderbilt was not received into the society of the village. As a rule people of wealth and prominence stopping at the village hotels receive invitations to parties from the Lincklaens, Ledyards, Ten Eycks, Burrs and many other aristocratic families of the place, but Mr Vanderbilt was ignored. Mrs. Morse, of New York [a Lincklaen-Ledyard], who was stopping at the Lincklaen House at the same time [as Vanderbilt was courting the woman he later married], was welcomed by the old families. It is thought that an inkling of Mr. Vanderbilt's object in visiting Cazenovia may have been received through her.

Such Anglophilic behavior became marked by the last years of the 19th century. Sidney T. Fairchild financed his purchase of the Cazenovia and Canastota Railroad with English capital, and his backers visited Cazenovia in 1878 (CR, 5–9–1878). George S. Ledyard and Charles Stebbins traveled to Utica in 1883 to attend a reception for Lord Chief Justice Coleridge (CR, 9–20–1883). In 1889 the Preston family entertained James Coats, who ran the American operation of the huge cotton thread business founded by his father, Sir Peter (CR, 6–6–1889). The Prestons then traveled with James and his fiancée, Mlle Marie Jean Adam (governess to the Preston children) to Scotland (CR, 7–11–1889). The Prestons hosted the Coats' wedding ceremony at their summer cottage, "Ormonde", later that year (CR, 9–19–1889).

A few years earlier the local paper had been moved to comment that (CR, 8–13–1885)

> It is always a pleasant custom to give appropriate names to country seats which are significant. Every English one, however simple, has its name. . . . The Lincklaen manor house has been called from its beginning "Lorenzo", "Willowbank" . . . , Mr. Fairchild's home, [is] named for the huge trees on the lake shore, brought from New Jersey in Mr. Lincklaen's saddlebags, more than half a century ago. . . . Mr. Preston's [actually Mr. Peet's] place will be called "Cloverly Lodge" and so recall Charles Kingsley's place. The Reverend Dr. Norton, who, from appearance will have an original and exquisite country seat, though costly, calls it "Notley–Mere". The "Notley" from the English county seat of Mrs. Norton's ancestors in Sussex, and "Mere" meaning "Lake".

In this same vein was Burr Wendell's organization of a "tilting tournament" in 1894 on the hilltop west of his house commanding a fine view of both village and lake (CR, 8–16–1894). Tilting tournaments had become high Tory events in England after the tournament at Eglington in 1840 (Girouard 1981, p. 88). Robert F. Hubbard enjoyed a very formal twenty–first birthday party in the English style (CR, 5–27–1897), and Charles Stebbins Fairchild flew the American flag at "Lorenzo" at half-mast upon the death of Queen Victoria (N. H. Kiley, personal communication, 1973). Fairchild was also a strong proponent of early intervention by the United States in World War I (New York *Herald*, 3–23–1917). Walter Oakman, husband of Anna Hubbard, took this a step further. In England at the declaration of war with Germany in 1939 he joined the Guards and served for the entire war with the British army (Mrs Walter Oakman, personal communication, 1984).

England's orderly, manicured landscape was greatly admired and, in one Cazenovia newspaper reports, linked by implication to the style of elite paternalist management that was coming to typify Cazenovia (CR, 11–5–1896).

> Buckland-on-the-Moor, a secluded village of Devonshire, England, has no public house, parson, policeman, or pauper. The squire owns all the land. The farms are small but profitable. The farm laborers live in the squire's cottages. When they fall sick the squire pays their wages as usual, and when they are too old to work any more they are continued on the pay list and potter about, doing what they please.

To a Ledyard Lincklaen or a Susan Fenimore Cooper of an earlier generation such behavior as the Cazenovia Republican was endorsing only 50 years later would have been bizarre and un-American. Cooper, for example, gave us a fine description of an American yeoman farm that almost defines the Puritan virtues of self-sufficient New England culture (1968, pp. 105–110).

> They kept four cows, formerly they had a much larger dairy, but our hostess had counted her three score and ten, and being the only woman in the house the dairy-work of four cows, she said, was as much as she could well attend to. One would think so, for she also did all the cooking, baking, washing, ironing, and cleaning for the family, consisting of three persons; besides a share of the sewing, knitting, and spinning. . . . The food of the family, as well as their clothing, was almost wholly the produce of their own farm; they dealt but little with either grocer or butchers. . . . The chief luxuries of the household were tea and coffee, both procured from the "stores", although it may be doubted if the tea ever saw China; if like much of that drink about the country, it was probably of farm growth also. . . . On one side stood a cherry bureau, upon this lay the Holy Bible, and that its sacred pages had been well studied, our friend's daily life could testify.

What made the re-Anglicization of the American elite possible from a practical standpoint was the improved quality of transatlantic travel after the Civil War. Better, safer, more reliable steamships removed the danger and rigor from the Atlantic crossing (Hugill 1986, p. 417). As early as 1888 the Syracuse Standard could comment that "over thirty of those who were last year members of The Owaghena [Club] are now in Europe" (8–8–1888). What made it necessary from a cultural perspective was the need of the old stock elite to separate themselves from the new immigrants of the cities. This re-Anglicization pushed a group that had previously acted relatively democratically more and more toward social differentiation by residential segregation and rural seclusion. Even in the countryside, where few immigrants were found in New York State, the new attitude came to prevail. A contemporary description of the rolling country around Cazenovia commented (Syracuse *Standard*, 8–8–1888):

> here the [summer] cottagers are wont at times to spend Sundays, high above all surroundings, and mayhap listening to the homely chat of an intelligent peasant who ekes out his honest livelihood upon the rocky hilltops.

By the turn of the century the old stock elite who moved there in the 1880s had made Cazenovia "home". In a telling brochure published under the title "An Ideal Spot for Idle Moments" by the "Summer Resort Committee" in 1905 (CR, 3–30–1905), Cazenovia was described as

> one of the oldest summer resorts in the state, and still retains all its natural grandeur, unspoiled by human hands. Its hills and valleys, forests and lakes, are so combined that each has its distinctive charm.

For many years it has been the resort of noted tourists who say it is surpassed by none abroad and has all the alluring features that entice travellers across the ocean for seclusion and rest.

The streets are broad avenues of crushed stone – compact and hard as asphalt pavements, untrammeled by the rush and roar of the electric car, which break the serenity of many resorts. Grand, stately old trees of every species, line these avenues on either side, their branches entertwining [*sic*] in shadowy archways, most inviting to the pedestrian. The social advantages are world famed, as the village society is made up of eminent people of wealth and culture, whose imposing and historic mansions are of colonial and modern architecture. The surrounding country has a park-like aspect and is dotted with many springs – all having medicinal properties.

The ultimate seal of approval was, however, set on Cazenovia in the re-Anglicization phase by the *Times* of London (CR, 9–29–1910).

The real summer resorts [of America] are of two classes, among which we may take Bar Harbor, Newport and Saratoga as typical on one hand, and Cazenovia on Lake Owaghena and Skaneatles, on the lake of the same name, on the other. The first are the haunts of millionaires – miniature Ostends, with even more formality of costume and entertainment – the glories of New York transported to another environment. . . . In the other class of summer resort . . . there is a great deal of informal entertaining, and it is at the club that it chiefly takes place. . . .

The keynote of the "summer colony" is enjoyment and not display. It is composed for the most part of splendidly healthy young men and maidens who find their chief delight in physical exercises, with the additional stimulus of competition and flirtation. The older members of the company play bridge or golf, fish, ride or drive, or watch their strenuous families competing at tennis, canoeing, bathing or some other sport.

For all this approval the Cazenovia of the Anglicized elite was not so much a home to its inhabitants as that of the republican period. Certainly it was not home in the sense that people's livings were earned in or around the village. Recently J. B. Jackson has argued for two views of America: a Jeffersonian America based on Renaissance values, strongly rooted in characterful places; and a rootless America, dominated by the highway, with interchangeable places (Jackson 1984, pp. 152–5). In the first community and home are propinquity. In the second communities are formed and re-formed at will. The republican northeast was a Jeffersonian place *par excellence*. Ledyard Lincklaen and Susan Fenimore Cooper praised Jeffersonian values above all others in their love of the artfully ordered picturesque, antiquity, and the natural world, however much their work was tinged with Anglophilia. After the Civil War the newly rich old stock elite expanded that Anglophilia toward anti-urban and anti-immigrant values. Yet their Anglophilia was improperly conceived. The English landed elite embraced some Jeffersonian values almost as fully as Jefferson himself, and continued to do so throughout the 19th and early 20th centuries by co-opting the new industrialists (Wiener 1981, p. 14). Their reason for so doing was that their prosperity and status were tied intimately to the land and its husbanding. The post-Civil War elite in America had much looser ties to the land, if any. Husbanding the land had nothing but symbolic value when wealth was derived from the sale of horseshoes, oil, or baking powder. This was an elite moving toward Jackson's second definition and away from the Jeffersonian ideal. For all the English airs adopted after the Civil War the most crucial from the English landed elite's point of view was lost. The country and the village ceased to be home in the fullest sense of the word, and became just a place to live. A pleasant place indeed, in a village like Cazenovia, but only one of many places for an elite with no deep roots in country life.

The American elite in the republican period was a true landed elite in the English model, its domestic roots deep in country life and with a love of the local and natural. It was based on class differentiation, but one sensitive to other groups and, indeed, the overall ecology. After the Civil War those roots were torn up, replaced by superficial pleasure in picturesque artifacts rather than in a complex ecology artfully modified, and by the importation of behaviors conceived elsewhere with little sympathy for the local in their application. It was the triumph of social differentiation at the price of lost sensitivity to class differences and ecological complexity. Sturdy farmers were transformed into peasants and robber barons were elevated to the nobility of chivalry. In the process home became something far more transient, and thus perhaps far less meaningful.

References

Published Materials

Fenimore Cooper, S. 1968. *Rural hours.* With an Introduction by David Jones. Syracuse, NY: Syracuse University Press. Reprint of 1887 edition.

Girouard, M. 1981. *The return to Camelot: chivalry and the English gentleman.* New Haven, CT: Yale University Press.

Grills, R. A. 1977. *Cazenovia: the story of a upland community.* Cazenovia, NY: Cazenovia Preservation Foundation.

Hawke, D. Freeman 1980. *John D: The Founding Father of the Rockefellers.* New York: Harper & Row.

Hugill, P. J. 1986. English landscape tastes in the United States. *Geographical Review* **76**, pp. 408–23.

Jackson, J. B. 1984. *Discovering the vernacular landscape.* New Haven, CT: Yale University Press.

Lincklaen, L. 1855. Rural objects in England and America. *Putnams Monthly* **31**, July, pp. 32–40.

Lockwood, A. 1981. *Passionate pilgrims. the American traveler in Great Britain, 1800–1914.* East Brunswick, NJ: Fairleigh Dickinson University Press.

Wiener, M. J. 1981. *English culture and the decline of the industrial spirit, 1850–1980.* New York: Cambridge University Press.

Other Materials

The *Cazenovia Republican*: a weekly newspaper (on microfilm).
Cazenovia Village Records, Cazenovia, NY.
Electrical World: a weekly trade magazine.
Niagara Mohawk Archives: archives of the Niagara Mohawk Power Corporation, Fulton, NY.
State Census of New York, County of Madison, 1875.
Summer Resort Committee, "An Ideal Spot for Idle Moments." (Cazenovia, NY: nd [1905]).

6

Social and symbolic places in Renaissance Venice and Florence

EDWARD MUIR & RONALD F. E. WEISSMAN

Introduction: Place as an historiographic problem

What sense have historians made of urban geography in the Italian Renaissance? The recent historiography of Renaissance Venice and Florence has given special emphasis to the sociology of space and geographically-based social ties. A sensitivity to place offers historians a rich context for the analysis of Renaissance society, and one which may change many common assumptions about the Renaissance social order because it runs counter to much of the traditional historiography concerning Italy during the Renaissance.

Any discussion of the social order and organization of urban Italy must begin with a recognition of the influence of a well-established view of the Renaissance city, accepted by Weberians, Marxists, and Burckhardtians alike. The traditional historiography commonly holds that the "cash-nexus" of Renaissance capitalism weakened traditional medieval geographic solidarities and clan loyalties, liberating the creative energies of the Italian townsman (Von Martin 1963, Goldthwaite 1980, Becker 1981). Left isolated without ties to neighborhood, family or guild, the townsman invested psychic energies in urban governments. The Renaissance, in this view, is nothing less than a paradigm of modernization, its urban history an exemplar of contemporary modernization theory.

According to this view which emphasizes the triumph of the rational and secular, space is typically assumed to have had a generally instrumental or functional value. Neighborhoods and other subdivisions of the city-state were no longer the sources of an influential man's power and had become subject to the superior controlling power of a new social, economic, and political order that transformed residents into citizens. Anticlericism and the spread of reason desacralized urban life so that only in the feudal countryside did medieval piety still dominate life. Historians subscribing to this view have long been impressed on the one hand with Italy's early form of urban capitalism and on the other with the flowering of republican politics. Thus, until recently, for most historians, the Italian city represented nascent modernity with all of its opportunities and dangers: rationalism, individualism, and the bureaucratic state.

In recent years, however, a different interpretation of Renaissance social geography has begun to emerge. In addition to that more traditional vision asserting the progressive weakening of place as a symbolic and social category of much importance, in the manner of Burckhardt (1958) and his more modern successors, the newer view, sensitive to the work of social geographers and anthropologists, makes "place," understood as the geographies of sociability and ritual, central to understanding Renaissance cities and Renaissance society.

This chapter examines the ways in which many historians currently view "the power of urban places" in the Italian Renaissance, focusing on the best documented and debated of cities, Florence and Venice. In particular, we will highlight two of the most important areas of inquiry in contemporary Renaissance scholarship: social geography, particularly networks of space-based sociability, and symbolic geography, the use of place to delineate, comment on, and transform the social order of the city.

The historical development of Italian cities through the Renaissance

With the possible exception of Flanders, Italy was the most urbanized area in the late medieval and Renaissance world. In 1550, more than 40 towns exceeded 10,000 in size; many were substantially larger, including Naples (60,000), Verona (50,000),and Rome (50,000). These figures do not do full justice to the vigor of late medieval Italian urbanization; in most cases, cities in 1550 were less populated, sometimes substantially so, than they had been 200 years before when the Black Death first struck.

From the 11th through 16th centuries, with the exceptions of towns serving principally as administrative centers, the major Italian cities were fundamentally commercial centers, operating as sea ports (Venice, Genoa, Pisa), key nodes of the European banking network (Florence, Siena), or centers of textile production (Florence, Lucca).

While the 11th century demographic explosion and the period's commercial opportunities aided Italian cities, urbanization was not a new phenomenon or one that had its origins in the medieval commercial revolution alone. At least 40 towns of Roman or pre-Roman ancestry had enjoyed an almost unbroken civic life since antiquity (e.g., Mantua, Bologna, Padua, Ravenna). These were, by the 13th century, in the minority. Nevertheless, the "new" urbanization of the high middle ages was clearly in part a product of existing urbanization, caused by movements of population from older, declining towns to new urban foundations, vigorous urban colonization of the Adriatic coast by the Venetians, and the demographic expansion of existing population centers (Jones 1974, Sestan 1977). A movement from countryside to town accounted for only part of Italy's urban growth.

Between the 13th and 16th centuries, most Italian towns to some greater or lesser extent experienced or were influenced by these developments:

(1) From the mid-11th through the 13th centuries, towns gained political independence from local bishops and imperial authorities, developed embryonic political institutions based on corporate groups such as families, guilds, and neighborhoods, and experienced explosive commercial and demographic growth. These towns were characterized by communal organization, that is, they were governed by shifting coalitions of private, essentially voluntary associations.

(2) By the end of the 13th century communes experienced a transformation and expansion in their authority as the commune developed from what had been, initially, a private to what was becoming a public, municipal body, a development accompanied by great factional violence as groups attempted to dominate, define, or resist that nascent "public" authority.

(3) During the 14th century communes stabilized political institutions around an oligarchy or lord, and reduced the level of political conflict. In Venice, the political order coalesced around a legally-defined patriciate at the end of the 13th century; in Florence, politics was conducted by a republican oligarchy consisting unofficially of a few hundred lineages and an inner core of a few dozen families.

(4) By the end of the 14th century, in part owing to the pressures of demographic catastrophe and disorder, cities developed permanent bureaucracies concerned with public order, the provisioning of foodstuffs, the protection and regulation of commercial assets, the regularization of public finance, and the control and ordering of public space.

(5) By the 15th century city-states had stabilized their external political relations as they expanded into larger territorial states, culminating in the supremacy of five centers of regional power: Milan, Florence, Venice, Rome, and Naples.

(6) During the 16th century Italian city-states underwent a thorough reordering of public life in the wake of foreign invasions, major epidemics, and economic contraction, leading to a much greater degree of centralized and planned authority in many spheres of urban organization, including the systematization of institutions, reducing overlapping and redundant medieval councils, greater regulation of charity, a reduction in the importance of older corporate forms of organization such as guilds, and a corresponding reordering of ecclesiastical authority around uniform institutions and mechanisms capable of ensuring orthodoxy, such as parishes.

(7) By the end of the 16th and early 17th centuries city-states suffered a significant relative decline in their economic importance as the Mediterranean itself gave up its primacy to Northwestern Europe. Italian cities lost their position as important entrepots in an international network of trade and became centers of regional trade and administration.

Renaissance civism

The outline of Italian urban history described above has led many historians to emphasize the civic character of the Renaissance Italian city. In so doing, they have accentuated the growth of the public sector, activist governments, public institutions, and civic consciousness. Since it tends to identify the components of Italy's apparent modern precocity, the notion of civism represents the most recent version of the traditional interpretation of the Italian Renaissance, although it has successfully avoided many of the anachronisms inherent in the Marxist, Burckhardtian, and Weberian views.

The historiography of civism began 30 years ago when Hans Baron (1966) described the evolution in Florence of a patriotic, republican, and modern body of political thought stimulated by threats from foreign tyrants. Baron's influential thesis gave a sense of coherence to the two principal fields of Italian Renaissance studies, urban history and the history of humanism. By linking the growth of a civic, republican consciousness among intellectuals to the political experience of the Florentines and the history of their relations with other states, Baron gave a common sense of order to both the literary and political history of the early Renaissance. Baron's work has prompted other scholars to see the creation of the bureaucratic and rationally-organized city-state as the most significant achievement of Renaissance Florence (Brucker 1977, Najemy 1982), and has also led to a revision of the history of Venetian humanist thought. (Bouwsma 1968, King 1986).

Although Florence and Venice have both been admired as paragons of Renaissance republicanism, the two civic worlds have also been contrasted in significant ways. Florence is typically viewed as an agitated and unstable yet introspectively creative polity and Venice, a serene, constant, conservative, and aristocratic republic (Brucker 1983; Bertelli *et al.* 1979-80). Viewing city politics in these ways, some scholarship has examined the ideological uses of civism. Fifteenth-century Florentine writers used the civic ideal as a weapon against disruptive elements in society (Najemy 1982); Venetian apologists adapted civism as a defense of the oligarchic status quo.

The "civic-mindedness" of Renaissance culture has been expanded by scholars well beyond humanist political thought and the development of the State. In the work of many scholars, including in some respects the present authors, the civic has thoroughly informed the nature of urban society and culture. Historians have discerned in Florence and Venice civic religion, civic ritual, civic painting – indeed, a whole civic world (Weinstein 1970, Becker 1974, Brucker 1977, Muir 1981, Weissman 1982, Patricia Brown 1984). In any discussion of social geography it is important to recognize how pervasive the civic model has become. It has developed into an all-encompassing view of Renaissance urban life, a view that emphasizes the triumph of the centralizing power of office-holding adult male patricians over the particularizing or potentially divisive forces of family, clan, neighborhood, clientage, status, gender, age, and, of course, place.

Whether one interprets civism as evidence of a broadly-rooted commitment to Renaissance republicanism (Baron 1966), as the hegemonic ideology of Renaissance elites (Najemy 1982), or as the self-interested rhetoric of humanist bureaucrats (Seigel 1966), the notion of a civic culture undoubtedly identifies much that is fundamental to the Renaissance, indeed to much Italian life more generally. As one anthropologist observed about Umbria, medieval and modern, even the smallest Italian communities considered themselves civilized and in possession of *civiltà* – the urbane influence of urban life (Silverman 1975). Nevertheless, during the last ten years, much historical research, even by some of those historians previously committed to the civic model, has begun to reveal the limits of civism as a core interpretive concept illuminating the workings of Italian Renaissance society. By concentrating even more systematically on the social components of Renaissance cities, rather than urban ideologies, historians have begun to recognize that civism may provide an incomplete view of urban life, and may even mask vital and divisive social processes under an ideological veneer of citizen unity.

To examine fully the importance of place in the Renaissance social order and to determine the extent to which place was one of those older solidarities rejected during the Renaissance in favor of a broader sense of civic allegiance, we turn our attention to the study of Renaissance social bonds, to sociability-relations as they were lived, not simply as they were imagined.

Social geography and sociability in Florence and Venice

Social relations in Florence and Venice followed much more complex patterns than one might predict from a narrow interpretation of the civic or modernization model (Weissman 1985). Although Florence was the birthplace of civic humanism, even Florentine patrician males were strongly tied to their neighborhoods, more dramatically so than the Venetian patricians. In both cities there was a significant disparity between the ideals of a civic rhetoric that spoke of the devotion of the citizen to the republican order, whose public spaces prominently dominated the city's center and the reality of urban social life with all of its elements of status, class, gender, and place.

The Renaissance civic-humanist bureaucrat, ever anxious to remake society in his own image and likeness, created a large urban planning literature, describing ideal cities and their socially useful and aesthetically perfect star-shaped, circular, or grid forms, having rationally-designed central places, spacious piazze, and wide boulevards (Simoncini 1974, Muratore 1975). The Florentine reality was quite different. Florence was polycentric, having many sources of social power and a physical structure with several distinct and dominant visual centers (Simoncini 1974, Goldthwaite 1980). Key institutions were dispersed throughout the city. Each quarter had one or more major mendicant churches or other churches sponsored by the Commune; in addition the major "public" towers dominating the skyline were widely separated, including those belonging to the baptistry/cathedral, the "town hall," the Palace of the Priors, the Badia, and the constabulary (the Bargello). Visually and conceptually, Florence's many-centered physical geography was chaotic and cluttered.

In the violent 13th century, each of Florence's many neighborhoods was dominated by one or more towers of its richest residents; several hundred of these spiny projections rose several stories above the surrounding buildings. Serving as urban fortresses during outbreaks of civil unrest, the towers were eventually demolished or reduced in size during the late 13th and early 14th century, as nascent public institutions attempted to control urban violence and factionalism. The towers were the most visible demonstration of the clan-centered culture in which kinsmen, friends, and retainers clustered together. Despite the destruction of the towers, a geographically-based clan culture continued throughout the Florentine Renaissance. Streets or piazze were as likely to be known popularly, if not officially, by the name of the most influential family residing there as by their formal names. Benedetto Dei's (1984, p. 79) 1472 description of the city lists Piazza de' Pitti; Piazza degli Strozzi; "the piazza of Santa Felicità, where the Barbadori and Chanigiani are;" "the church of San Iacopo (Sopr'Arno) situated among the Ridolfi;" a street "which goes from the Guidetti [residence] as far as Luca Capponi." These names were instantly familiar to the Florentine, whose cognitive map of the city was drawn with finely detailed shadings of power, influence, and turf. To Dei the city was the sum of its parts, an agglomeration of 50 piazze, each with "churches, houses and palaces of the principal citizens of the reggimento, merchants and shops catering to many needs."

Just as Florence presents a polycentric cityscape, Venice fails to conform completely to the civic-humanist ideal of a rationalized urban plan, but the city's center comes closer than does Florence to the humanists' architectural ideals. Long before the Renaissance humanists wrote, Venice had concentrated its most prestigious and powerful institutions in a centralized place anchored by the Doge's Palace and adjacent basilica of Saint Mark. The Doge's Palace housed the doge (Venice's head of state), all of the elected councils, and most of the bureaucratic offices of the republic. Next door one of the great churches of Christendom harbored the relics of an evangelist and dominated the vast ballroom–like piazza that spread in front of it. The city's principal financial officers, the Procurators of Saint Mark, occupied the buildings around the piazza, buildings whose regular, classicizing facades emphasized the solid reasonableness of Venetian institutions. The Library of Saint Mark, the Mint, and a second church completed this vast complex whose long building history juxtaposed diverse styles that sensitive architects had unified through iconographic echoes, the use of similar building materials, and quotations from earlier architectural forms (Demus 1960, Sinding-Larson 1974, Pincus 1976, Wolters 1983, Tafuri 1984).

Most travelers to Venice arrived, quite literally, at the center of the city. Ships anchored in a basin in front of the Doge's Palace, and passengers disembarked at the Molo where two pillars topped with statues representing the protector-saints Theodore and Mark demarcated the entrance to the city. One experienced Venice from the center outward, Florence through a multiplicity of possible paths. Venice lacked the mainland tradition of urban fortresses and tower societies, and although some minor alleys and embankments retained clan names,none of the principal places did. Unlike Benedetto Dei, Marin Sanudo (1980) described his city as a series of civic and ecclesiastical monuments, and the only time he attached clan names to neighborhoods was in his account of noble houses that had died out (pp. 178-80). Venetian patricians strived to find a housing site along the Grand Canal, the location whose prestige outweighed any advantages gained by remaining in and identifying with a specific neighborhood.

In the early medieval stage of lagoon reclamation, Venice evolved as neighborhoods expanded creating what was for many centuries a multicentered city composed of distinct island communities. Venice grew by private development around two spatial forms, the *campo-rio* (square-embankment) complex and the courtyard, both of which emerged through the patronage of a few clans who followed their own priorities without an overall conception of the city. Even the exact course of the Grand Canal came to be defined only haphazardly as land reclamation and new structures spread outward from the campi (Dorigo 1983, pp. 492–502, Bellavitis & Romanelli 1985, pp. 37–9). Although private in origin, the *campo* soon became the center of a neighborhood and was collectively used, whereas the second form, the courtyard, provided coresidence around a wealthy patron and was, therefore, exclusive though not always private.

In the 13th century two parallel developments began to reduce the political significance of *campi* and courtyards as the constituent elements of patrician-dominated neighborhoods. On the one hand, the Venetian state became the closed preserve of a patrician caste whose exclusive access to political office was ratified by the closing of the Great Council in 1297. The patricians of the Great Council no longer based their power on a body of retainers or neighborhood clients but on their ability to influence the balloting of their patrician peers. On the other hand, adult siblings and different generations of the same family dispersed their residences about the city (Lanfranchi 1955, pp. xiii–xix; cf.Bellavitis & Romanelli 1985, pp. 49–50). The semi-feudal enclaves of the great families broke up, in part because of the growing power of the state and in part because a changed economy lessened the desirability of such enclaves. As a result the ability of patricians to control their neighborhoods or even the courtyards of their own houses slipped away. Politics in Venice became unlike politics in Florence as patricians, entitled by birth to political rights, sought to make contacts and to build careers by means of election to office. New men rarely entered the ranks of the closed political class. In Florence new men could achieve office by allying themselves with a powerful neighborhood clan, and old families sustained themselves by maintaining local contacts.

In both cities, however, outside of the political class, most people continued to find their most vital daily contacts in their piazza-based neighborhoods. These neighborhoods did not conform to the stereotype of the medieval city in which members of the same craft lived close together in the same district. Most neighborhoods were socially heterogeneous, containing both the palaces of the rich and the tenements of the poor, and members of many different trades. With a few exceptions production was organized on such a small scale that practitioners of the same trade had no particular incentive to live in close proximity to one another. Apart from ethnic ghettoes of foreign workers, residential segregation was normal only for the artisans in a few specialized crafts. In Florence, for example, several large neighborhoods housed clusters of the city's wool workers at various points along the city's rural/urban periphery: Sant'Ambrogio, Santa Lucia sul Prato, San Frediano, Santa Maria in Verzaia (Wyrobisz 1965, pp. 307–43, Brucker 1977, Cohn 1980, Weissman 1982). The artisans in dangerous and noxious crafts were isolated by law or custom, such as Venice's glassblowers who built their furnaces on Murano or the Venetian tanners who settled on Giudecca. In Venice the government-sponsored Arsenal, with the intensive concentration of workers' residences in its environs, provided the most significant industrial exception to the pattern of occupational integration, and the only completely homogeneous neighborhood was the peripheral parish of the Nicolotti fishermen (Zago 1982, pp. 12–16). In both cities most artisans lived among and married the daughters of craftsmen in other professions, although in the 15th century, members of the Florentine working class may have experienced a higher rate of parish, if not occupational, endogamy than did their Venetian brothers (Cohn 1980, Romano 1987).

Florence and Venice, despite their 15th-century populations in excess of 40,000 and 120,000 respectively, were for most people intimate societies of face-to-face bonds. Renaissance capitalism did not create the impersonal, socially segregated society that one associates with modern urbanization and economic development. Despite the exceptions, such as the Venetian patriciate, that have led traditional historiography astray, the basic units of Italian social organization in most towns remained, throughout the Renaissance, family and neighborhood. The division of Italian towns into neighborhoods in which families and their retainers congregated was, of course, as old as the Roman republic (Mengozzi 1973). Indeed, as more than one historian has suggested for the Florentines, the social world of most Italians was centered around, *"parenti, amici, e vicini"* – kin, friends, and neighbors (Klapisch-Zuber 1976).

Burckhardt wrote that the Renaissance townsman was freed from obligations to older clan and corporate solidarities, and others, seconding Burckhardt, have suggested that such transformations were occasioned by the rise of capitalism and civism and the demise of feudalism. The most recent scholarship, however, has demonstrated that such solidarities, far from collapsing, persisted well into the Early Modern period (Brucker 1977, F. W.Kent 1977, D. Kent 1978). The extended family , although it lost its 13th-century corporate status as Tuscan tower society (*consorteria*) or Genoese fictive clan (*albergo*), remained a fundamental social unit, serving as the organizing force behind Italian commerce (F. W. Kent 1977), qualifying one for guild membership and membership in other corporate groups (Doren 1940), continuing as an essential component of prestige, and influencing one's honor, status, and ability to participate in urban politics (Chojnacki 1973, Kent 1978, D. Finlay 1980, Najemy 1982).

For Florentines kinship was inextricably linked to the neighborhoods in which kinsmen clustered. The *gonfalone*, with its civic responsibilities, its identification with clan and faction, and its theatrical, public, face-to-face character, stood as a bridge between the particularist forces of the medieval commune and the emerging civic world of the more recognizably modern state. The neighborhood *gonfalon*'s role was ambiguous, at once a nucleus in miniature of the larger republican culture, and, at the same time, a focus for local loyalties and the furthering of local interests.

The most intimate neighborhood unit was, of course, the parish. The Florentine parishes had, in the 13th and early 14th centuries, served as the smallest units of civil administration, resolving quarrels, for example, and the next largest division, the ward or *gonfalone* constituted Florence's militia, the term *gonfalone* deriving from military pennants emblazoned with a totemic animal or heraldic figure symbolizing the ward: Red Lion, Green Dragon, Ladder, and Viper. Florence was repartitioned in the mid-14th century. The four civil quarters of the town, named for four of the major churches that dominated the life of those quarters, were themselves subdivided into fourths. These subdivisions, retaining the name of *gonfaloni*, each encompassed two or more parishes. Operating as formal corporations, the larger, secular *gonfaloni* replaced the parishes as the principal nuclei of civic order. The Florentine neighborhoods that clan factions used as military power bases in the 13th century had become by the end of the 14th century bases of political clientage (D. Kent 1978, Molho 1979).

A recent study of one of these neighborhoods during the 15th century by Kent & Kent (1982) has highlighted the social, political, and administrative importance of the *gonfaloni*. Florentines qualified for political office through their neighborhoods, because scrutinies of all politically-eligible males were conducted by neighborhood committees headed by the *gonfalon*'s chief officer, its standard bearer. An honorable reputation among and cultivation of neighbors, particularly of the leading families exercising patronage in the district, were essential for political success.

Reputations were won and lost on a neighborhood stage, in full view of one's intimates. Fortunes were also unmade in "public view," for each ward was assessed its share of the city's tax burden. *Gonfalon* committees estimated the net worth of each resident of the neighborhood, based as much on hearsay, gossip, and personal observation as on formal declarations, and tax payments would be negotiated further when assessments came due (Kent & Kent 1982). Given the highly personal nature of this process, it is no surprise that members of the politically-eligible class clearly had strong incentives to maintain neighborhood ties, to avoid moving out of one's ancestral district, and to remain among kin and friends within the confines of a narrow area (Brucker 1976, F. W. Kent 1977, D. Kent 1978, Weissman 1982).

Neighborhood could also generate strong animosities and jealousies, for the piazza served as a common stage bringing together a Florentine's many social selves: kinsman, friend, political ally, tax assessor, business partner, client, parishioner. Managing them and maintaining numerous loyalties was an arduous task. The neighborhood was, indeed, a stage upon which honor could be won or lost in the politics of everyday life (Kent & Kent 1982, Weissman 1982).

Historians have rightly stressed the corporate character of Florence's sixteen *gonfaloni*, but the formal legal boundaries of the *gonfaloni* were more important for the interests of the mature males of the patriciate than for others whose primary relationships were not political. "Neighborhood" or *vicinanza* was often informal, located in the affective ties that linked members of the same parish or neighbors bordering the same piazza. Social relations as varied as those of marriage, political sponsorship, the witnessing of legal instruments, friendship (as measured by sponsorship for membership in lay religious associations) and godparenthood frequently clustered within the neighborhood, loosely defined. Neighborhood patrons helped arrange low-cost housing, provided loans, assisted with tax assessments and payments, wrote letters of recommendation, offered political backing, produced administrative appointments and employment, and introduced clients into networks of friends (Heers 1974, Klapisch-Zuber 1976, Hughes 1977, Molho 1979, Cohn 1980, Kent & Kent 1982, Weissman 1982).

The specific role of neighborhood varied by class, by status, by age, and almost certainly, by sex. For the Florentine citizens who were eligible politically and wealthy enough to pay taxes, the *gonfaloni* and quarters of the town had significant meaning. It was, after all, under the banner of the *gonfalon* that each male citizen assembled under threat of fines during the city's chief civic pageant, the feast day of Saint John the Baptist. For the socially marginal – the poor and the working classes, adolescents, and women – neighborhood was more fluid and amorphous, and could include piazza, street corner, or alley. Most commonly, however, lower class sociability coalesced around the parish, and in the 14th century and again in the late 15th century the *popolo minuto* organized neighborhood festive bands which staged mock and occasionally real turf battles during feast days (Trexler 1980, Weissman 1982).

The role of neighborhood varied temporally as well. Although both were important during the 13th and early 14th centuries, the fortunes of Florence's parishes and *gonfaloni* appear to have varied inversely thereafter. By the 14th century Florentine parishes lost formal civic administrative responsibilities to the *gonfaloni* and festive identity both to the more prestigious and more spiritually attractive mendicant churches which organized each quarter of the city, and to citywide confraternities (Trexler 1980, Weissman 1985). As Florentine republicanism declined during the latter half of the century, and crumbled during the first quarter of the 16th century, the *gonfaloni* lost their political relevance. Under Medici potentates and finally, under Medici dukes, political status derived from loyalty to the Medici family, not to a strong neighborhood retinue. At the same time, however, neighborhoods and parishes became increasingly important as centers of working-class and middle-class sociability as revealed by marriage patterns (Cohn 1980). By the middle of the 16th century, the parish, newly energized by the forces of Catholic reform, was the only remaining source of corporate solidarity, in the wake of the collapse of *gonfaloni* and guilds (Weissman 1985).

Venice exhibited weaker forms of neighborhood organization than did Florence. A high level of residential mobility within Venice, evident by the 13th century, contributed to the weakness of the parish as the locus for identity formation. As we have noted, males from the notable families pursued greater rewards and influence by competing for civic offices and seeking government favors, so that in Venice the pattern of patronage was far more city-wide and far less localized than in Florence or Genoa. In Venice, however, patronage may have been peculiarly sex and class specific. Dennis Romano (1987) has suggested that Venetian patrician women, in contrast to their husbands, developed well-articulated local patronage networks largely because women were secluded in their palaces and seldom appeared in public beyond the parish confines. Romano (1987) has found evidence that lower-class women in the 14th century chose a patrician woman from their own parish to act as executor of their wills whereas lower-class men almost never designated a male patrician to serve in this delicate capacity. In Florence, on the other hand, artisans might be obliged to look to a local protector for a job, favor, or loan. A particularly revealing comparison can be found in the fact that festive turf battles in Florence were between neighborhoods whereas in Venice the organized lower-class fights over the possession of a bridge were between the two largest divisions of the city, the Nicolotti and the Castellani, groups comprised of workers from opposite sides of the Grand Canal. The hard core of the combatants in Venice, moreover, seems to have come from those two exceptionally homogeneous worker neighborhoods, the fishermen Nicolotti and the Arsenal workers of Castello.

In the face of general noble indifference to neighborhood and parish affairs, better-off commoners performed the service functions for Venetians that might in Florence fall to the hands of the local padrone. Venetian commoners, for example, dominated the parish-level priesthood. Their influence came not so much from their role as spiritual confessors and advisors and even less from their abilities as preachers but from their involvement in the secular world. Parish priests served as executors of wills, exercised the power of attorney, acted as notaries, invested in commercial ventures, and were particularly valued as sources for small loans (Romano 1987). However, neither priest nor parish figured pre-eminently in the social world of the male Venetian, although the research suggesting this conclusion for Venice is at a far more primitive state than is the work on Florence. The parish may have been an important spiritual locus for women, but men in Venice as in many Renaissance cities experienced the sacred by joining a city-wide confraternity or by acting as a lay patron for a monastery or mendicant church (Pullan 1971, Weissman 1985, Goffen 1986).

On a different level many functions such as the organization and drilling of militia members, policing, the distribution of tax assessments, dispersal of grain to the needy, collecting of census information, and managing routine maintenance of streets and bridges fell to lay officials at both the parish and district level, in Florence the *gonfaloni* and quarters, in Venice the *contrade* and *sestieri*. In Florence these jobs seem to have provided local power brokers, in their capacity as "amateur" citizen politicians, opportunities to exercise patronage, but in Venice they were the responsibility of minor and insignificant bureaucrats or patricians far outside of the inner circle of rulers. The one area where patricians took their neighborhood responsibilities seriously was in serving time as policemen, called the Lords of the Nightwatch, but these posts were exclusively manned by poor or youthful patricians (Ruggiero 1980, pp. 14–17, Pavan 1981b, pp. 342–3). The give and take of Venetian elections and the exclusive, oligarchic character of the Council of Ten, dramatically lessened the importance of patrician patronage of the neighborhoods.

In many respects Venice appears to be a better centralized, more bureaucratic, less particularist city – in short the more "modern." Explaining the differences between these two cities in light of modernization theory, however, would be a mistake. The peculiar urban and administrative history of Venice, an experience that emphasized co-operative work, centralized planning, regulation of building, systematic reclamation, and the political power of capital accumulation, was largely the product of the city's special environmental situation. The lagoons created by the Po, Adige, Brenta, and Piave rivers have since classical times produced a delicate environment that has required recurrent and systematic intervention in order to permit human habitation.

Sensitivity to collective survival sustained in Venice a concern for public utility maintained at the expense of private property rights and of corporate groups unable to muster the resources necessary to dredge canals, reclaim swamps, prevent the accumulation of sandbars, and build sea walls. All canals, channels, ponds, alleys, and even courtyards eventually became public property available for common use (Strina 1985). Permanent magistracies watched over all aspects of the lagoon's aquatic environment creating a tradition of careful, centralized management unknown and unnecessary elsewhere (cf. Pavan 1981a, Crouzet-Pavan 1982). Until its conquest of the mainland, in fact, many of Venice's early military adventures against Padua resulted from the rival city's upstream diversions of the Brenta and Noventa rivers, diversions that threatened the delicate balances of the lagoon. Venice's environmental imperative fundamentally altered its social life, granting to the civic-minded patricians of the central government a special legitimacy no Florentine magistrate ever achieved. For those outside the political class, neighborhood and parish life retained a certain vitality in Venice, but for the patriciate real power lay not in the neighborhood campo but in the Piazza San Marco.

The symbolic geography of the Renaissance city

Power in Venice's Piazza San Marco or in Florence's Piazza della Signoria was as much sacred as it was secular. The ideal of the medieval city as a sacred community, a mystical corpus, was, of course, central to the political ideology of many medieval cities. One only needs to recall late medieval paintings of Italian cities cradled in the arms or hands of a saint or bishop, as San Gimignano cradled his city in Taddeo di Bartolo's painting, to see demonstrated the deeply perceived connections between cities and the sacred in the psychology of late medieval and Renaissance Italy.

The inhabitants of Florence and Venice conceived of their own cities rather more as enchanted communities, mystical bodies, alive with symbolic meaning than the architectural, institutional, and political entities historians have often made them by reason of the surviving evidence of buildings, monuments, and archival documents. Socialization began in early childhood to teach the meanings conveyed through names, emblems of all sorts, civic myths, and collective rituals (Trexler 1980, pp. 9–128). Names and symbols were attached to urban places – streets, piazze, churches, and town halls – and public rituals always took advantage of a symbolic geography of the city bringing a special meaning to urban spaces.

In the symbolic geography of Italian Renaissance cities one might distinguish between "place" seen primarily as a cultural artifact, embedded in a grid of other meaningful objects and locales, and "space" understood as a physical location that related to other locations. Places are spaces with names, spaces with evocative, multidimensional identities (cf. Tuan 1977, Cosgrove 1982). In Italian cities place names flower everywhere giving urban spaces life, meaning, and emotive identity and remaining for centuries as fossilized remains of past lives while the spaces themselves are refurbished, demolished, enlarged, or abandoned. The courts, houses, squares, walls, suburbs, streets, churches, and monastic enclaves of Italian cities function within a complex of interrelationships, but each also has a certain architectural autonomy, that gift of spatial construction, which allows the function to change independently of pre-existing connections with other spaces. Places, however, with their self-perpetuating names are subject to different variables, those cultural processes that bequeath meaning to human experience. Although there is always some connection between a place and its space, they can often change quite independently of each other.

In many Italian cities places represented certain values that transcended their precise spatial function. For Florentines to enter the "piazza" or for Venetians to go to the Rialto meant far more than a jaunt to the town square or market place. These names signified a whole sector of human activity with an appropriate dress, gestures, and decorum: for Florentines the piazza referred to the town hall and participation in communal politics with the corresponding elevation of the civic over the parochial and for Venetians the Rialto signified the bargaining, buying, selling, speculating, and maximizing of personal profits we would call capitalism and they called tending to their affairs. The semantic range implied by these labels reached far beyond the urban spaces they denoted.

Contemporaries attuned to the semantic richness of place names employed them in defining a person's character and status because accomplished citizen-actors communicated messages by following certain patterns as they moved about the city's spaces. One of early 15th-century Florence's wealthiest citizens, Palla Strozzi, for example, sought to avoid envy:

He always avoided publicity. He never went into the Piazza except when he was sent for, nor into the Mercato Nuovo. In going to the Piazza, so as to avoid observation, he would go by Santa Trinità and turn into the Sant' Apostolo as far as the Via Messer Bivigliano. Then he would reach the Piazza and enter the Palazzo without delay (Vespasiano 1963, pp. 236-7).

For Palla and his contemporaries, the "Piazza" meant city hall, a whole complex of political ambiguities, a mix of patronage, civic rectitude, envy, and honor, a place with a very public and well-populated square, where petitioners, clients, and opponents lay in waiting.

The signifying power of the city's places was most dramatically magnified by the touch of the sacred. But where, precisely, was the locus of the sacred in the Renaissance city? Whether or not the sacred could be localized in space became, after all, a major issue in the theological conflict between Catholics and Protestants, the former insisting on the divine presence in the Eucharist and treating relics as special objects of devotion, the latter refusing to acknowledge such an impious mixing of spirit and matter (Davis 1981). The dispute, after all, was never purely theological. Relations with the sacred provide an idealized pattern of earthly social relations (Christian 1972), and changes in attitudes toward the sacred altered the means by which Renaissance townsmen and women might form their social identities.

In Catholic cities the sacred occupied space in two ways. One was a theologically legitimate and precise attribution of divinity to the Host and holiness to relics, objects that were for all practical purposes mobile. Peter Brown has argued (Brown 1981, pp. 86-105) that one of the distinguishing characteristics of early Christianity was its belief in the mobility of the sacred. Christians replaced sacred wells, caves, and trees with Christ's Eucharistic body and the corpses of martyrs for the faith, objects which could be moved from place to place. Churches, therefore, were not so much sacred spaces as shelters for holy objects. In the medieval and Renaissance periods believers also often attributed an aura of sanctity to painted or sculpted images, particularly to pictures of the Virgin. When approaching various sacred objects reverence was conveyed by the performance of prescribed gestures (Baxandall 1972, Barasch 1976, Trexler 1980), in short a demonstration that one has been properly socialized.

Sacred objects and the behaviors they evoked were understood to have beneficial social effects. Peace pacts between feuding families included the swearing of an oath in the presence of the Host. To calm a riotous crowd priests would proceed throughout the city with a miracle-working image or relic. Churches, of course, were consecrated to harbor holy relics and the divine Host, but the sacredness of a church was essentially a moveable space, the hallowed zone around an object that was usually housed in a consecrated building and framed on a fixed altar but that could be moved when necessary. When it was moved the hallowed zone went with it.

Richard Trexler's pathbreaking work on Renaissance ritual behavior provides a useful and needed model for a serious rethinking of the social function and operation of ecclesiastical architecture and other sacred spaces.

Churches and monasteries were holy because of the objects they contained and the activities they permitted: "The place does not sanctify the man but the man the place" (Francesco da Barberino, quoted in Trexler 1980, p. 52). The impulse to decorate and embellish churches and especially altars may have come from a certain anxiety about the mobility of the sacred. A saint who was ill-treated or forced to dwell in shabby surroundings might just allow his or her body to be "translated" elsewhere. And the theft of relics was always a danger. Many of Venice's most important relics including the body of Saint Mark and the head of Saint George had, in fact, been stolen in North Africa or the Near East and brought to Venice by traveling merchants and crusaders (Geary 1978, Muir 1981, pp. 78–102).

The legitimate (as opposed to furtive) method for moving a sacred object was an ecclesiastical procession. Such processions accompanied the celebration of major religious holy days, the onset of social crisis such as drought, famine, or plague, or the celebration or commemoration of a great military victory. Even on that most frequent of occasions, when a priest rushed to administer extreme unction to a dying sinner, he was to be accompanied by an acolyte ringing a bell to warn passers-by of the approach of the body of Christ. More elaborate processions of clerics, whole parishes, confraternities, or government officials transformed public streets into temporary *viae sacrae*.

Despite the temporary communitas produced by such processions, the transformation and the resulting behavioral effects, exemplified by social peacemaking among citizen spectators and participants, were only temporary. During the 14th century the Venetian Council of Ten tried to prolong the sacred's pacifying effect on street behavior by having sacred images with frames called *capitelli* erected on the outside walls of houses and churches. It would appear that the Council believed that it would be more difficult to continue street fights if every hundred odd paces or so a picture of the Madonna forced would-be combatants to stop and show reverence. Little holy places, therefore, dotted the urban map, so that churches and monasteries by no means constituted the only fonts of the holy in an otherwise secular city. Hallowed places, moreover, had to be identified by signs, specifically by the cross, which located for the viewer the divine objects within a church. Saint Bernardino of Siena recognized the power of such signs when he recommended that crosses be placed on the walls of the city to restrain urination in public (Trexler 1980, p. 54). Holiness welled up like so many bubbling fountains in Venice and Florence to provide townsmen with numerous daily opportunities for contact with the sacred (cf. Christian 1972).

The images, objects and their signs that flooded the city contributed to the second way the sacred occupied space. Because of certain rituals or actions a building or location could retain a residual holiness; as consecrated buildings, set apart for the celebration of the sacraments, churches effused a specialness that for many believers was independent of the relics contained there. The subtle emanation felt to come from ecclesiastical buildings merged with the aesthetic response to architectural beauty. It may have been impossible for most persons to distinguish between the emotions evoked by visual delights, music, and incense on the one hand and by the spiritual aura of the saints on the other.

The residual specialness of certain places could, as well, come from sources other than the sacred. Just as there could be a generalized spatial holiness, so too there were locations polluted by crime or evil. After the murder of Giuliano de' Medici at the hands of the Pazzi conspirators during Mass, Florence's cathedral had to undergo an elaborate ritual purgation. In many Italian cities, executioners routinely returned a murderer to the scene of the crime where the removing of the criminal's hand or the gouging out of an eye cleansed the location of the residual evil (Ruggiero 1980, pp. 47–49, Muir 1981, pp. 245–52, Edgerton 1985, pp. 126–64). In the most notorious cases, especially those involving treason or rebellion, the guilty party's house was torn down, the earth salted, and a defamatory memorial erected on the spot. A particularly gruesome example can be found in 1585 after a revolt in Naples where, on the location of the rebel leader's destroyed house, officials erected an altar upon which 23 severed heads were placed. This act of justice itself came to be seen as perpetuating the pollution rather than cleansing it; after many months the archbishop petitioned to have the altar and heads removed so that the place could be properly purified (Mutinelli 1856, vol. 2, pp. 140-58). Ever insecure about the stability of their regimes, the Florentines were particularly aware of the need to isolate the attendant profanation of capital punishment from the hallowed city hall and removed their gallows outside the city walls (Edgerton 1985, pp. 139-41). Places, therefore, could be made holy or profane through human activities, the memory of which gave certain locations special qualities that could only be altered through a ritual act or the passage of time.

The locations which best established and solemnized order also became the most likely settings for displays of disorder and scenes for pollution. Riots and revolts in early modern cities often followed a decidedly ritualistic pattern, in which alternative banners, blazons, or insignias replaced the symbols of degraded power; the routes of rebellious demonstrations followed the paths of sacred processions; and blasphemies mocked the formulas and gestures of respect (cf. Davis 1975, pp. 152- 87). Mass demonstrations employed a language of space that can only be translated by reference to the symbolic geography of the city (Burke 1983, Muir 1984). Where a victim was lynched might be just as important as who he was and what he had done, as can be seen by the Medici partisans' preoccupation during the turmoil after the Pazzi conspiracy with finding the most degrading locations to display their victims' dismembered bodies (Poliziano 1978) or in the 1216 vendetta between the Amidei and Buondelmonti, the apocryphal origin of the Guelph and Ghibelline controversy, whose bloody end occurred in full view of the Ponte Vecchio at the spot where the original insult initiating the cycle of revenge had itself been committed.

Less dramatically evil yet still profane activities regularly penetrated consecrated spaces, creating ambiguity about the range of behavior acceptable in the presence of the sacred. The separation of the sacred from the corruption of business activity required of Christians by Christ's casting of the money-changers from the temple was weak in Renaissance Italy where the market needed churches to facilitate business and where religious behavior was often merely another form of negotiation. A market might best be understood as a special place where a peculiar kind of limited trust could govern transactions (Becker 1981). Without such an atmosphere of trust the extension of credit and the mechanisms of business activity were not possible. Ever suspicious of others, Florentines and Venetians sought to sanctify their commercial dealings (Trexler 1980, pp. 111–12, 263–70). Contracts were customarily notarized, signed, and witnessed in a church where the presence of the Host and holy relics might invest the parties with a fear of divine punishment for breaking their word. Proximity to a church, therefore, was just as necessary for a market space as ready access to transportation routes. One of the oldest standing churches in Venice for example, is in the center of the Rialto market. Such profane uses for churches provoked protests from reforming preachers, such as Bernardino of Siena, but until the Counter Reformation, clerics had little success in convincing the laity to remove business deals, parades of prospective brides, club meetings, and political assemblies from sacred spaces. Sacred places, however, never quite matched the aura of relics and the Host. Even Saint Bernardino argued that a sacrilege against a holy object was far worse than one simply perpetrated within a holy place (Trexler 1980, pp. 53-4).

Burial practices are even more revealing of how sacred spaces configured the urban landscape. Florentines and Venetians, over the objection of the parish clergy, were free to choose their burial place, a right which meant that individuals and families might favor some church other than their parish, a tendency reinforced, at least in Venice, by high levels of residential mobility. Tombs crowded around favored altars. The richest citizens either appropriated the high altar for themselves or built their own chapels along the walls of a preferred church in order to guarantee their family permanent proximity to the divine and to establish a relationship of perceived reciprocal protection with a spiritual power such as an onomastic saint or the Madonna. In Venice most family chapels are in parish churches, but a significant minority in both cities sought special spiritual benefits outside of their parish and paid for tombs in convents, monasteries, or mendicant churches often located at some distance from the family house or palace. In Florence, virtually all churches constructed during the Renaissance were built according to a plan lining the interior with private chapels (Goldthwaite 1980). Florence's strong sense of neighborhood and parish identity are, of course, exemplified by Medici patronage of their neighborhood church, San Lorenzo, patronage which they guarded so jealously that other neighborhood families were discouraged from contributing to its decoration (Kent & Kent 1982).

As a counterweight to parish loyalties Florentine city fathers promoted civism via the cult of Saint John the Baptist, whose popularity spread from the Romanesque baptistry where all of Florence went to be baptized. The baptistry and the adjacent cathedral became the spiritual center of Florence and the beginning and end for most processions. In addition government buildings, especially the city hall, represented political salvation through the display of sacred signs and symbols. A raised platform in front of

Florence's hall became an altar during public ceremonies thereby directly imputing divine sanction to public authority (Trexler 1980, p. 49).

The more fully centralized civic buildings of Venice especially intrigued provincials, ever anxious to emulate the styles of the powerful. All across the mainland provincial nobles imitated Venetian styles in their palaces as if the urban appearance and imperial power of Venice were somehow intermixed. After they conquered mainland towns, Venetians set up statues or reliefs of the lion of Saint Mark as insignia of their rule. But the relationship between political domination and the magnetic power of the San Marco complex had even stronger correlates. In town after town, such as in Vicenza and Udine, Venetian governors refashioned the main town square into an imitation of the great piazza and piazzetta of the capital, exemplifying power over space by the physical imitation of a particular place. The Venetian patricians' dominion over foreign cities came to be a mere extension of the model by which communal domination had been illustrated within the city. Loyalty might best be promoted, these provincial squares seemed to say, by copying Venice's symbolic centerpiece because certain appearances encouraged civic responsibility.

Despite the visual and rhetorical propaganda created by these two regimes, the civic always held a tenuous and vulnerable position among competing forces, especially before the middle of the 16th century and especially in Florence. The significance of the symbolic core of the city was always more important and more meaningful for the males of the patriciate than it was for women, youths, disenfranchised artisans, immigrants, and foreigners. In Florence, the mobility of the sacred and the city's annual liturgical cycle conspired to give every major neighborhood and that neighborhood's chief lay patrons, a chance to demonstrate its charisma to the civic community, a chance to link the collective honor of its inhabitants to the process of showing devotion to the city's chief saints. And so, in 1472, Florentines celebrated the first great feast of the year, San Giovanni, in the baptistry, proceeded to celebrate Corpus Christi in the cathedral, moved across the Arno to celebrate the Annunciation in San Felice in Piazza, the Ascension in the Carmine, the feast of Saint Augustine in Santo Spirito, and celebrated other major feasts in San Lorenzo, Sant' Anna, the oratory of San Onofrio, San Marco, San Giorgio, and Santa Mariade' Servi (Dei 1984, p. 93). In a similar pattern the Venetian doge and Signoria attended special Masses throughout the city for the various annual feasts although Venice was more likely than Florence to commemorate in its civic liturgy historical events important for the city (Muir 1981, pp. 212-23).

In the Florentine celebration of the feast of the Magi the link between space, sacred charisma, and earthly honor was particularly obvious to even the casual observer. In the Medici-sponsored feast of the "republican" 15th century, representatives of each of three quarters of Florence dressed as Magi kings, paid homage to the fourth quarter, passing the Medici palace and proceeding to "Bethlehem," the Medici-dominated convent of San Marco, to adore the Christ Child (Hatfield 1970, Trexler 1980, pp. 424–45).

The ritual recognition of neighborhood churches and monasteries differed in Venice from the situation in Florence in one distinctive way. After the late 14th century, neither the parishes, *sestieri*, nor any of the other neighborhood divisions of the city were ever represented in a Venetian ritual the way the quarters of Florence were honored in the feasts of the Magi and San Giovanni. The constituent elements of the Corpus Christi rite in Venice, for example, were corporate groups, especially the confraternities, and the greatest annual festival, the marriage of the doge to the sea, depicted the city as a whole in a mystical union with its watery environment (Muir 1981, pp. 119-34). The routes of Florentine and Venetian processions followed the cities' symbolic geography, creating a formal commentary on urban space through a recurrent public drama.

Conclusion

The Renaissance city witnessed the spectacular growth of the "public," as a political concept, as a way of transferring social trust from persons to institutions, as a moral concept promising impartial justice, and as a set of claims on authority and allegiance which transcended individual loyalties. Neither in Florence nor in Venice, however, was the authority of the "public," or its claim on a citizen's allegiance, honor, obligation, and trust accepted unambiguously.

Medieval Italian urban institutions arose in the communal period amidst competition for power among private organizations, many of them geographically based, but by the late 13th and 14th centuries cities developed a strong "civic," public sphere. Despite a growing faith in public institutions evident from the 13th century, another organizational culture coexisted uneasily with the growth of the civic world. This culture, known commonly as *clientela* or *amicizia* – patronage – emphasized loyalty to persons rather than abstract principles. Italian patronage derived in part from feudalism, but equally from the urban organization of Roman antiquity, and persisted well beyond the Renaissance into the modern period. An ambivalent trust in impersonal institutions, a tendency noted in republics as well as princely regimes to organize parties and factions around great families and personal networks of dependents, and an ambiguous boundary between "public" and "private" were all characteristics of this culture of friendship and *clientela*. Local loyalties, those of neighborhood, kinship, and friendship, and the bonds of patronage, stronger in Florence, perhaps, but not entirely absent in Venice, always competed with loyalties and obligations to public institutions.

Florence and Venice represent different ways of resolving that conflict between center and periphery characteristic of Renaissance cities. Elites competed to dominate sacred spaces by patronage of churches and chapels, placement of arms, and sponsorship of feasts. Given a culture in which strongly-rooted traditions of civic obligation and private loyalties coexisted, it is not surprising that there remained competing conceptions about the proper uses of sacred spaces for private and public purposes, conceptions that reveal the elasticity of meanings and differing notions of decorum within the same culture.

Venice displayed a more precise hierarchy of sacred and profane spaces, a time-bound, sometimes inverted, occasionally subverted hierarchy, but nevertheless an essentially political scheme that symbolically organized much of the urban plan. In Florence, however, the relative strength of private power ensured that private groups would compete with public authority by elevating their private spaces to the symbolic status achieved by public power.

This chapter has examined two related manifestations of the "power" of place to orient social relations and to convey meanings in Renaissance cities. The goal of the public control of space was to influence the loyalties and obligations of persons. To accomplish this the sacred was employed, on occasion, to work a miraculous restructuring of social obligations in a way impossible merely through the expansion of public domination over urban spaces. By expanding the sacred from a civic center both Florentine and Venetian republics sought to advance the claims of the public at the expense of private ties and obligations. This advance was measured by the ability of the public to control the use of spaces and to regulate representations of the sacred.

Florentines, in particular, needed the protection of the sacred in order to act in a world in which all obligations, all relations, were viewed as corrupting, particularistic, and potentially conflictual. But if the sacred could be used to enhance the public domain and to engender loyalty and trust to the republic, so too could the manipulation of sacred spaces and places be used by private groups to enhance their charisma and their claims. Thus, even, perhaps especially, the sacred was subject to the same particularist forces as was the secular. One of the constants of Florentine history is that every regime laid claim to legitimacy by employing the city's vocabulary of sacred space.

In Florence, social ties to local places constrained the thoroughgoing expansion of public over private space and civic over particularist loyalty. In Venice, the necessity of controlling and dominating a difficult habitat, that ever difficult space that would disappear into the sea without constant intervention, led to the subordination of geographic-based loyalties in the necessarily civic interest of collective ecologic survival. The criteria for the appropriate use of the sacred was also more stable in Venice. Saint Mark had a greater, more unifying and more lasting hold on Venetian loyalties than was the case in Florence where Saint John the Baptist had to compete with the Medici Magi and living prophets such as Savonarola. The civic triumphed in Venice, not completely, perhaps, but completely enough to allow central-place institutions to dominate the Venetian social and spatial order.

For both the Florentine and the Venetian, the sacred, through the language of space, permeated the secular world, filled it with meaning; in Florence the sacred order, symbolized by sacred spaces, attempted to remedy the natural defects and limitations of the secular, civic order. For the Venetian, the interpenetration of sacred and secular places, relations, and meanings, as with the marriage of the sea, confirmed the authority of Saint Mark and the rightness of the ordering of the Venetian's social and ecological world. The Florentine language of space was a discourse about the competing claims of public and private; in Venice that "language" celebrated the successful subordination of private to public.

References

Barasch, M. 1976. *Gestures of despair in medieval and early Renaissance art*. New York: New York University Press.

Baron, H. 1966. *The crisis of the early Italian Renaissance: civic humanism and republican liberty in an age of classicism and tyranny*, New edition. Princeton: Princeton University Press.

Baxandall, M. 1972. *Painting and experience in fifteenth-century Italy: a primer in the social history of pictorial style*. Oxford: Clarendon Press.

Becker, M. B. 1967–68. *Florence in transition*. 2 vols. Baltimore: Johns Hopkins University Press

Becker, M. B. 1974. Aspects of lay piety in early Renaissance Florence, in *The pursuit of holiness*, C. Trinkaus (ed.) Leiden: E.J. Brill.

Becker, M. B. 1981. *Medieval Italy: constraints and creativity*. Bloomington: Indiana University Press, 1981.

Bellavitis, G. & G. Romanelli 1985. *Le città nella storia d'Italia: Venezia*. Bari: Editori Laterza.

Bertelli, S., N. Rubinstein, & C. H. Smyth (eds) 1979–80. *Florence and Venice: comparisons and relations. Vol. 1: Quattrocento, vol.2: Cinquecento*. Florence: La Nuova Italia.

Bouwsma, W. J. 1968. *Venice and the defense of republican liberty: Renaissance values in the age of the counter reformation*. Berkeley & Los Angeles: University of California Press.

Brown, P. 1981. *The cult of the saints: its rise and function in Latin christianity*. Chicago: The University of Chicago Press.

Brown, P. F. 1984. Painting and history in Renaissance Venice. *Art History* **7**, pp. 263–94.

Brucker, G. 1976. *The civic world of early Renaissance Florence*. Princeton: Princeton University Press.

Brucker, G. 1983. Tales of two cities: Florence and Venice in the Renaissance. *American Historical Review* **88**, pp. 599–616.

Burckhardt, J. 1958. *The civilization of the Renaissance in Italy*. 2 vols. New York: Harper & Row.

Burke, P. 1983. The Virgin of the Carmine and the revolt of Masaniello. *Past and Present* **99**, pp. 3–21.

Chojnacki, S. 1973. In search of the Venetian Patriciate: families and factions in the 14th century. In *Renaissance Venice*, J. R. Hale (ed.), pp. 47–90. London: Faber & Faber.

Christian, W. 1972. *Person and God in a Spanish valley*. New York: Academic Press.

Cohn, S. K. 1980. *The laboring classes in Renaissance Florence*. New York: Academic Press.

Cosgrove, D. 1982. The myth and the stones of Venice: an historical geography of a symbolic landscape. *Journal of Historical Geography* **8**, pp. 145–69.

Crouzet-Pavan, E. 1982. Murano à la fin du Moyen Age: spécificité ou intégration dans l'espace vénitien? *Revue historique* **268**, pp. 45–92.

Davis, N. Z. 1975. *Society and culture in early modern France*. Stanford: Stanford University Press.

Davis, N. Z. 1981. The sacred and the body social in sixteenth-century Lyon. *Past and Present* **90**, pp. 40–70.

Dei, B. 1984. *La cronica*, Roberto Barducci (ed.). Florence: Francesco Papafava Editore

Demus, O. 1960. *The Church of San Marco in Venice: history, architecture, sculpture*. Dumbarton Oaks Studies, no. 6. Washington:Dumbarton Oaks.

Doren, A. 1940. *Le arti Fiorentine.* 2 vols. Florence: Le Monnier.

Dorigo, W. 1983. *Venezia: origini.* 2 vols. Milan: Electa.

Edgerton, S. Y. 1985. *Pictures and punishment: art and criminal prosecution during the Florentine Renaissance.* Ithaca and London: Cornell University Press.

Finlay, R. 1980. *Politics in Renaissance Venice.* New Brunswick: Rutgers University Press.

Geary, P. J. 1978. *Furta sacra: thefts of relics in the central Middle Ages.* Princeton: Princeton University Press.

Goffen, R. 1986. *Piety and patronage in Renaissance Venice: Bellini, Titian, and the Franciscans.* New Haven and London: Yale University Press.

Goldthwaite, R. 1980. *The building of Renaissance Florence: an economic and social history.* Baltimore: The Johns Hopkins University Press

Hatfield, R. 1970. The compagnia de' Magi. *Journal of the Warburg and Courtauld Institutes* **23**, pp. 107–161.

Heers, J. 1974. *Le clan familial au Moyen Age: étude sur les structures politiques et sociales des milieux urbains.* Paris: Presses Universitaires de France.

Hughes, D. O. 1977. Kinsmen and neighbors in medieval Genoa. In *The medieval city,* Harry A Miskimin *et al.* (eds), pp. 95–111. New Haven: Yale University Press.

Jones, P. 1974. La storia economica: dalla caduta dell'Impero Romano al secolo XIV. In *Storia d'Italia,* vol. 2, pp. 1469–810. Turin: Einaudi.

Kent, D. 1978. *The rise of the Medici faction in Florence.* Oxford: Oxford University Press

Kent, D. & F. W. Kent 1982. *Neighbours and neighbourhood in Renaissance Florence: The district of the Red Lion in the fifteenth century.* New York: J. Augustin.

Kent, F. W. 1977. *Household and lineage in Renaissance Florence: the family life of the Capponi, Ginori, and Rucellai.* Princeton: Princeton University Press.

King, M. L. 1986. *Venetian humanism in an age of patrician dominance.* Princeton: Princeton University Press.

Klapisch-Zuber, C. 1976. Parenti, amici, vicini, *Quaderni Storici* **33**, pp. 953–82.

Lane, F. C. 1973. *Venice: a maritime republic.* Baltimore and London: The Johns Hopkins University Press.

Lanfranchi, L. 1955. *Famiglia zusto.* Venice: Comitato per la pubblicazione delle fonti relative alla storia di Venezia.

Lanfranchi, L. & G. G. Zille 1958. Il territorio del ducato veneziano dall' VIII al XII secolo. In *Storia di Venezia,* vol. 1, pp. 3–65. Venice: Centro Internazionale delle Arti e del Costume.

Martines, L. 1979. *Power and imagination: city states in Renaissance Italy.* New York: Alfred A. Knopf.

Mengozzi, G. 1973 (1933). *La città italiana nell' alto Medio Evo.* Florence: La Nuova Italia.

Molho, A. 1979. Cosimo de' Medici: Pater Patriae or Padrino? *Stanford Italian Review* **I**, pp. 5–33.

Muir, E. 1981. *Civic ritual in Renaissance Venice.* Princeton: Princeton University Press.

Muir, E. 1984. The cannibals of Renaissance Italy. *Syracuse Scholar* **5**, (2), pp. 5–14.

Muratore, G. 1975. *La Città Rinascimentale: tipi e modelli attraverso i trattati.* Milan: Gabriele Mazotta.

Mutinelli, F. 1856. *Storia arcana ed aneddotica d'Italia raccontata dai veneti ambasciatori.* Venice: P. Naratovich.

Najemy, J. M. 1982. *Corporatism and consensus in Florentine electoral politics, 1280-1400.* Chapel Hill: University of North Carolina Press.

Pavan, E. 1981a. Imaginaire et politique: Venise et la mort à la fin du Moyen Age. *Melanges del'EcoleFrancisedeRome(MoyenAge–TempsModernes)***93**, (2), pp. 467–93.

Pavan, E. 1981b. Recherches sur la nuit venitienne à la fin du Moyen Age. *Journal of Medieval History* **7**, pp. 339–56.

Pincus, D. 1976. *The Arco Foscari: the building of a triumphal gateway in fifteenth century Venice*. New York: Garland.

Poliziano, A. 1978. The Pazzi conspiracy. Translated by Elizabeth B. Wells in *The earthly republic: Italian humanists on government and society*, B. Kohl & R. Witt (eds), pp. 305–22. Philadelphia: University of Pennsylvania Press.

Pullan, B. 1971. *Rich and Poor in Renaissance Venice: the social institutions of a Catholic state, to 1620*. Oxford: Basil Blackwell.

Romano, D. 1987. *Patricians and Popolani: the social foundations of the Venetian Renaissance state*. Baltimore: The Johns Hopkins University Press.

Ruggiero, G. 1980. *Violence in early Renaissance Venice*. New Brunswick: Rutgers University Press.

Ruggiero, G. 1985. *The boundaries of Eros: sex crime and sexuality in Renaissance Venice*. New York and Oxford: Oxford University Press.

Sanudo, M. 1980. *De origine, situ et magistratibus urbis venetae ovvero La città di Venetia (1493–1530)*, A. C. Arico (ed.). Milan: Cisalpino–La Goliardica.

Seigel, J. 1966. 'Civic humanism' or Ciceronian rhetoric? *Past and Present* **34**, pp. 5–48.

Sestan, E. 1977. La città comunale italiana dei secoli XI–XIII nelle sue note caratteristiche rispetto al movimento comunale europeo. In *Forme di potere e struttura sociale in Italia nel Medioevo*, G. Rosetti (ed.). Bologna: Il Mulino.

Silverman, S. 1975. *Three bells of civilization: the life of an Italian hill town*. New York: Columbia University Press.

Simoncini, G. 1974. *Città e società nel Rinascimento*, 2 vols. Turni: Einaudi.

Sinding-Larsen, S. 1974. *Christ in the Council hall: studies in the religious iconography of the Venetian republic. Acta ad archaeologiam et artium historiam pertinentia*, no. 5. Rome: Institutum Romanum Norvegiae.

Strina, B. L. 1985. *Codex publicorum (codice del piovego)*. Venice: Fonti per la storia di Venezia.

Tafuri, M. 1984. *"Renovatio urbis": Venezia nell' eta di Andrea Gritti (1523–1538)*. Rome: Officina Edizioni.

Trexler, R. C. 1980. *Public life in Renaissance Florence*. New York: Academic Press.

Tuan, Yi-Fu 1977. *Space and place: the perspective of experience*. Minneapolis: University of Minnesota Press.

Vespasiano da Bisticci 1963. *Vite di uomini illustri del secolo XV*, W. George & E. Waters, translators (Renaissance princes, popes, and prelates). New York: Harper & Row.

Von Martin, A. 1963. *Sociology of the Renaissance*. New York: Harper & Row

Wolters, W. 1983. *Der bilderschmuck des Dogenpalastes in Venedig*. Steiner.

Weinstein, D. 1970. *Savonarola and Florence: prophecy and patriotism in the Renaissance*. Princeton: Princeton University Press.

Weissman, R. F. E. 1982. *Ritual brotherhood in Renaissance Florence*. New York: Academic Press.

Weissman, R. R. E. 1985. Reconstructing Renaissance sociology: the 'Chicago School' and the study of Renaissance society. In *Persons in groups: social behavior as identity formation in Medieval and Renaissance Europe*, Richard Trexler (ed.). Binghamton: SUNY Press.

Wyrobisz, A. 1965. L'attivita edilizia a Venezia nel XIV e XV secolo. *Studi Veneziani* **3**, pp. 307–43.

Zago, R. 1982. *I Nicolotti: storia di una comunita di pescatori a Venezia nell'eta moderna*. Abano Terme: Francisci Editore.

7

Power and place in the Venetian territories

DENIS COSGROVE

Place as an organizing concept owes its popularity in contemporary human geography to its incorporation of human meaning into otherwise abstract and mathematical spatial concepts (Relph 1976, Eyles 1985). Places are physical locations imbued with human meaning. This is not to say that they have to be inhabited, merely that they must possess significance for people. Thus the North Pole is a place, while Longitude 76°W, Latitude 43°N is not, it does not resonate with social significance, at least not until we are told it is Syracuse, NY. Landscape, which is the subject of this paper, has also enjoyed a renewed popularity among geographers seeking a more humane social science. Landscape differs from place in a number of ways, the least important of which is that in common usage it implies a larger space. It implies a different kind of relationship between people and location. The fullest relationship we can have with a place, it is often claimed, it to live in it, to be a true insider. This explains in part the focus among humanistic geographers on everyday life. For landscape we are always an outsider, because landscape is something seen, viewed from beyond it (Cosgrove 1984). Landscapes are not landscapes for insiders, they are home, place of work, or recreation, or banishment (Berger 1976). This fact of landscape is implicit in its unshakeable *visual* connotation, the pictorial sense with which it first enters the English language and which has never been erased (Cosgrove 1985). Landscapes are pictorial *images* and their history is entirely bound up with the inscription of environmental images by various media on various surfaces: painted on canvas, sketched on paper, photographed on film, gardened onto the earth's surface. A geography of landscape is a geography of images, a study of ways of seeing and representing (Cosgrove & Daniels 1988).

To claim that landscape is a way of seeing the world, of patterning, composing, framing, and organizing space into an image demands that we enquire what is brought to the world in order to create the image. Clearly more is implied by the question than the empirical issues of media and technique. Landscape images create and transform or reconstitute space to suit human ideas of order, truth, beauty, harmony, fear, and so on. In this sense landscapes have a moral dimension, they speak to notions of how the world *should* be, or more accurately how it should *appear* to be. Once we move into this moral sphere we recognize that ideas of

104

scientific absolutes and empirical certainty carry little warranty. Images are statements, constantly open to challenge, alteration, reinterpretation, endless further glosses. This is why the analogy of landscape to text is such a fertile one (Samuels 1979). And the most interesting questions concerning images have less to do with the heritage of particular images, with their specific form or structure, the media of their inscription (although these are vital stages in interpreting them) than with their contextual meaning. Landscape is no exception. As a cultural geographer I seek to interpret the meaning of a landscape as it constitutes and articulates social and environmental relationships. And landscape images do play a significant role in such relationships, whether it be in massaging the harsh realities of agrarian change in the 18th-century English countryside (Barrell 1980, Solkin 1982), projecting the imperialist aspirations of the young American Republic (Novak 1980), or promoting the Futurist vision of 20th-century urban society (Gold 1985). Social relations inevitably incorporate relations of power and authority, and this dimension is by no means absent from landscape images, their creation, endurance, and meaning. Most landscape images that survive have been produced for the powerful and they articulate social and technical power in some measure (Harley 1988). Indeed the landscape way of seeing implies authority, at least over space. This is revealed in the phrases in which landscape is commonly used: we "command" a view for example. Landscape images control pictorial space as their patrons in general controlled terrestrial space. Landscape images order the pictured environment and staffage as their patrons have ordered the historical environment and others who dwell within it. The relationship between image and history is not of course a simple one and I do not wish to imply an ideological symmetry between them. But within the theme of this collection, power and place, it is worth reminding ourselves of the power of the image and the place of landscape in the exercise of that power. Here I shall illustrate the interaction of landscape and power in a detailed examination of a set of images that are more than merely an appendage to a place but which constitute an integral part of the composition and meaning of that place.

Landscapes of the Villa Godi

Covering the internal walls of a late Renaissance villa in the foothills of the Alps some 15 miles north of the city of Vicenza and some 65 miles west of Venice is a fascinating series of painted frescoes (Fig. 1). They represent views over landscapes, combining images of classical Greek and Roman architecture with topography, scenes of everyday life, and vernacular structures clearly taken from the local area. They were painted in the early 1550s by local painters (Gualtiero Padovano, Giambattista Zelotti, and Battista del Moro) (Rossi 1982). They are the earliest examples of landscape frescoes covering the walls of a Palladian villa and represent a genre which would reach its highest development in the celebrated cycle of the 1560s painted by Paolo Veronese at the Villa Maser. I do not wish

Figure 7.1 Detail of internal fresco, mid-16th century, Villa Godi-Malinverni, Lonedo di Lugo, Vicenza.

to comment on the provenance of these works, still less to pass judgment on their style or quality. I intend simply to examine carefully what they represent and the context in which they were produced, because I believe that these pictures can reveal much to us about the relationship between power and place as it was conceived and experienced in the world of a powerful patrician family in the Venetian land territories during the second half of the 16th century. The approach I wish to adopt could perhaps be termed *iconographic* in the sense that it attempts to disclose, by close examination of a graphic image, and of the contextual documents which surround it, the meaning of the work and its place in the cultural and social world in which it was produced. The approach is interpretative; I myself am constructing a story, an account of power and place in the 16th-century Venetian landscape.

The landscape frescoes of the Villa Godi at Lonedo di Lugo are but one element of an internal decoration which includes allegories of art, *virtù*, intelligence, and doctrine, and references to Classical culture, literature, and history. Many of the human figures represented are in the mannerist style of gigantic stature, some of them mythical, others historical. And the painters have incorporated representations of 16th-century patricians involved into this historical pageantry: drawing aside curtains to reveal the scene, sitting within a casement or holding back a hound from entering the pictorial narrative. The whole decorative scheme consciously occludes the distinction between the world of the mid-16th century and the classical and mythical world of the paintings. This is specifically true of the landscape scenes. They are illusionistic works, framed by painted windows whose designs accurately reflect the panelling of the real windows giving over the landscapes of the Astico valley at Lonedo. In some rooms it is possible to stand facing the window framing the actual landscape and turn 180 degrees to view the illusory scene through a mirror image of the aperture. Foreground elements like trees are often partly obscured by the frame to suggest reality. The eyeline is the same as that seen from the rooms of the villa and the line of vision is led along a serpentine path formed by overlapping bands of color to a horizon at the depth of the picture. The atmosphere is a subdued morning or evening light emphasized by the chalky texture of the medium. In the foreground are figures: walking or seated in the arcadian landscape, sometimes in classical garb, sometimes in 16th-century costume. In the middle ground is water, either a river or a coastal bay across which we see the white columns and pediments of classical architectural structures. The literary reference is to the coastal villas described by the Classical Roman writers, Pliny or Cicero and celebrated in late Renaissance villa literature (Crescentio 1561, Gallo 1565, Sartori 1981), but the buildings are generalized rather than specific. Beyond rise mountains, blued in the distance, but whose forms echo those of the Dolomitic Alps that stand to the north of the Veneto and whose representation had become standard in the works of Venetian landscapists from Bellini and Giorgione in the early years of the century. But these are not purely Classical scenes; they generally incorporate vernacular buildings and activities obviously recorded from the cinquecento countryside around the villa: watermills, thatched peasant houses, farmsteads, and fishermen. These elements serve to suggest that the Classical world has been resurrected within the contemporary world of the late Renaissance Veneto and that the patricians who occupy the villa can access that earlier world at will. The fields and working landscape around the villa are brought into the frame just as the Classical Virgilian landscape is imaginatively projected outwards to the world beyond its walls. The success of the illusion depends upon the *trompe l'oeil* of perspective, the calculated mathematical technique that formed the foundation of Renaissance art.

This is the surface meaning of the landscape decoration of the villa, a decorative genre that derives not from Venice but from the *terraferma* city of Padua and which Renato Cevese has traced to the writings of Alvise Cornaro and Giovanni Maria Falconetto, Paduan humanists of the late Renaissance and realized above all by Gualtiero Padovano as a quite separate tradition of landscape art from that of the city of Venice itself (Cevese 1970). It was a genre that would become almost standard in the Venetian villas of the second half of the century. It appears at first sight merely an aristocratic conceit, a *jeu d'esprit* and it can and has been studied in terms of a long tradition of summer villa decoration learned from Classical authors and recommended in many of the texts published in 16th-century Venice as handbooks for the numerous wealthy landowners then improving their estates and constructing summer residences on them for relaxation from the city and for supervising the running of their agricultural enterprises (Puppi 1972, Sartori 1981). The significance of these illusions however is much broader than this. I have argued elsewhere that the relationship between the framed landscape way of seeing the world, controlling and appropriating it visually, and the exercise of actual power over land and production is a close one historically (Cosgrove 1984). Landscape acts as a visual signifier of place and its cultural construction. I believe that this is true of the illusory landscapes of the Venetian villas. They combine with the architecture of the villa, its relations to the surrounding countryside, and the activities that went on there to reinforce a particular social and environmental relationship, a relationship of power being exercised with increasing intensity over people and land, power expressed directly in economic and environmental exploitation, in cultural activities and in the construction of places.

The architecture of the Villa Godi

The Villa Godi (Fig. 2) was commissioned by Girolamo Godi in 1540 and completed by 1542 as the inscription above the door makes clear: "Hieronymous Godi Enrici Filius Fecit Anno MDXLII". Its architect was Andrea Palladio, at this time in his early thirties and well established as the architectural voice of the patrician class in Vicenza, an important and wealthy subject city on the Venetian *terraferma*. Palladio, originally a stonemason from Padua, had been given a humanist education in the Vicentine academy of Giangiorgio Trissino, a Florentine educated humanist from one of the richest and most influential families in Vicenza. His academy had been the training school of the aristocratic youth of Vicenza in the late 1520s and 1530s providing a broad humanist curriculum of ancient history, Greek, Latin, mathematics, geography, rhetoric, and physical education (Wittkower 1962, Ackerman 1966). It served to impart a highly exclusive culture to the young patrician class which, although no doubt for most merely a veneer over their traditional more mundane pursuits, developed the self-image of physical and intellectual stature, of *virtù* which set those who had received it apart from mere *mechanicians*, the non-citizen artisans and guildsmen of Vicenza (Ventura 1964). In the 1540s and 1550s this generation of humanist educated noblemen was coming into its inheritance and stamping the mark of its self-image onto the urban and rural landscapes of Vicenza.

Figure 7.2   Villa Godi-Malinverni, Lonedo di Lugo, Vicenza by A. Palladio.

Palladio, their chosen architect, was no mere mechanician. Not only had he received the same humanist education as his patrons, he practiced what Renaissance humanists regarded as the noblest creative profession: *architecture*, the "queen" of the arts, which called upon the broadest possible skill and wisdom (Barbieri 1983). Palladio's incorporation into the aristocratic world of provincial Vicenza as the person most able to articulate its world view had been signified in the mid 1530s by his selection by the city's Noble Council as architect of the double set of loggias to surround the *Palazzo dei Signori*, the council chamber of Vicenza, its rival to the Doge's Palace of Venice, and its central political icon (Barbieri 1970). Girolamo Godi had, as member of a Council committee, supervised Palladio's work on the copula of the cathedral at Vicenza and the two men shared the same cultural world.

The Villa Godi was the first country house that Palladio designed, although as Rudolf Wittkower has pointed out it "contains the germs of all the elements of Palladio's future developments" (Wittkower 1962, pp.

70–71). Here are the elevated central residential block with its aristocratic pediment standing above the storage spaces, the subordinate flanking loggias which embrace the landscape, and the internal plan of symmetrically ordered rooms proportioned according to a sophisticated mathematical conception of harmonic proportions, a conception which called for the most detailed understanding and manipulation of geometry. As an early conception the Villa Godi still shows stylistic signs of the traditional closed defensive structures adopted by signorial houses in the *terraferma*. The full openness to the landscape seen for example in the Villa Emo at Fanzolo or in the Villa Barbaro is not so apparent here. But the visual relationship between the villa and the surrounding countryside is critical to its conception. The Villa Godi has always attracted the admiration of its visitors for its positioning in the landscape. In 1680 Filippo Pigafetta in his description of the Province of Vicenza referred to:

> . . . Lonedo with the famous palace of the Counts Alessandro and Hieronimo Godi which commands this landscape, of graceful Greek and Latin architecture, complementing the hills and filling the valleys, with marble steps, and loggias, hall, suites of rooms, paintings, masterful frescoes, and splendidly furnished. . . . (Pigafetta 1974),

while Andrea Scato in 1649 referred to the "fertile and most delicious" hills in which the villa sits (Mazzi). A more recent writer points out that while the villa is architecturally not one of Palladio's most mature and inspired conceptions, nevertheless,

> it owes its principal beauty to the smiling landscape in which it rises, constituting, in the rolling topography of the hills, a point of harmony between the broad and generous verticals of the great trees, and the wide horizontals of the countryside, its classical forms reduced to hints, and dominating the broad scenography of nature. The villa is, like others of its kind, immersed in the green of pastures and of high trees which emphasise it and give it the solemnity of classical peace. (Gingio 1967)

It is quite clear that Palladio himself was highly sensitive to the landscape potential of the site. A signorial house had stood here earlier and its location is central to the largest consolidated land parcel in the Godi estate (Fig. 3). But the architectural organization of the building makes maximum use of the aesthetic qualities of the site, as Palladio makes clear:

> At Lonedo, a place in the Vicentine, is the following fabrick, belonging to Signor Girolamo de Godi. It is placed upon a hill that has a beautiful prospect, and near a river that serves for a fish pond. To make this place commodious for the use of a villa, courts have been made, and roads upon vaults, at no small expense. The fabrick in the middle is for the habitation of the master, and of the family. The master's rooms have their floor thirteen feet higher from the ground,

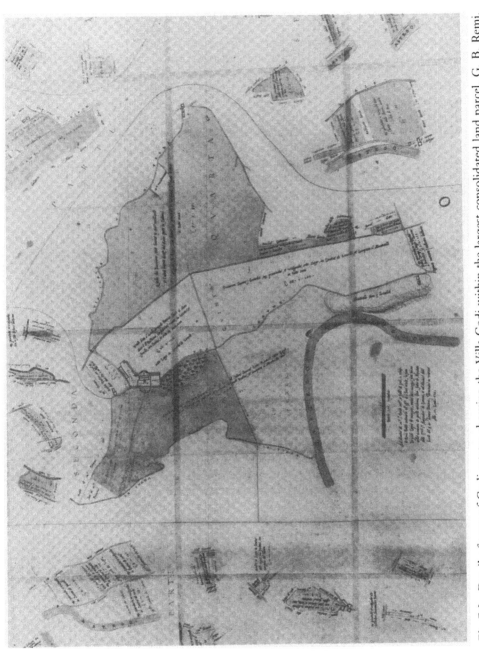

Fig 7.3 Detail of map of Godi estates showing the Villa Godi within the largest consolidated land parcel. G. B. Remi. 14 July 1579 (Biblioteca Bartoliana, Vicenza, ZXVI Lonedo).

and are with ceilings; over these are the granaries, and in the part underneath . . . are disposed cellars, the places to make wines, the kitchens and such other places. (Palladio 1965)

Such an arrangement not only served the practical purpose of separating the functional uses of the villa as agricultural center from its role as a place of retreat for a noble family during the summer months, but it maximized the inherently scenographic quality of the site by offering long views over the landscape from the elevated living quarters. This arrangement is most obvious from the central entrance to the villa which is approached via a dramatic flight of steps leading to a tripartite entrance lobby, a covered open-air room, a space that mediates inside and outside and offers spectacular perspectives over the Astico valley and the lines of hills beyond. This lobby opens into a central hall running the depth of the villa and lit by an enormous "Serlian" window which gives views over the orchard at the rear of the building. On either side of the main hall open the living apartments with their frescoed walls. Like its central pediment, the loggias shown on the plan in Palladio's *Four Books of Architecture* as flanking the residential block and designed to house agricultural produce, equipment, and housing for overseers and servants, were never built.

The architecture of the villa, like the frescoes that decorate its internal walls, insists on its aesthetic relationship with the surrounding countryside, while at the same time serving efficiently the practical needs of an estate center. Aesthetically and functionally the design clarifies a distinction between the decorated *piano nobile* of the central block of the villa, a place of refined human dignity, of art and passive relaxation away from the public demands of the urban palace, and the upper and lower floors and flanking loggias which serve functional estate requirements. The family apartments occupy the nexus of theoretically proportioned architectural elements (Wittkower 1962, pp. 126–41), just as the villa as a whole is the hub of an articulated pattern of working farms and fields. A balance of integration and distinction between aesthetics and production is achieved architecturally as it is in the paintings by the use of geometric proportion: it is a mathematical conception. I shall return to the significance of mathematics and proportion below, but first I wish to consider the social and economic context of the Villa Godi.

The Godi family and its estates

The Godi family was one of the wealthiest and most powerful patrician clans in 16th-century Vicenza. Its origins are obscure, but in 1510 the family held seven seats on the Noble Council of Vicenza, placing them among the 40 most influential clans of the 300 who had citizenship in this city of 30,000 people (Rumor 1899, pp. 90–1). The head of the family in the early years of the century was Enrico-Antonio (1451–1536). He was a member of the Guild of Lawyers whose membership was open only to the most important patrician families. He had fought on the side of Venice during the Wars of

Cambrai in the first decade of the century when the city was occupied by Imperial forces. In the course of his 85 years Enrico-Antonio amassed considerable lands in three areas of the province of Vicenza, largely by a process of debt peonage. Peasants were loaned monies by urban patricians like the Godi for the purposes of paying state taxes which were heavily weighted against the rural areas by a provincial council dominated by urban interests (Corazzol 1979). When these loans, secured against peasant land, fell due and could not be met the land reverted to the creditor who then either rented it back to its former owner or, if it lay adjacent to existing land parcels in the noble's estate, consolidated it for commercial farming with wage labor. By this method both Venetian families seeking more secure financial returns than were available from maritime trading and provincial urban patricians were dispossessing the peasants and small landowners of the *terraferma* throughout the 16th century (Ventura 1968).

Enrico-Antonio also owned a large house in the center of Vicenza on the site of the present Prefecture, valued in 1563 at 2000 ducats, in the highest class of urban real estate estimates in that year (Battilotti 1980, p. 124). In his will Enrico-Antonio divided his estate between three of his sons (Mantese 1966–67, BCB 1536). The oldest Girolamo who commissioned the villa at Lonedo in 1540–42, died without issue in 1544. The second, Marcantonio, had only daughters whose parts, upon marriage and dowry payment reverted to the male line so that in 1572 the entire estate fell into the hands of Pietro Godi and was left on his death in 1575 to his sons in seven equal parts to be passed on through the male line (BCB 1573). The exact arrangements made by the Godi brothers for dividing their inheritance are unknown to us, and partly for a very good reason, a reason however that has also provided us with a unique and precious record of the exact size and pattern of the landed estate of a major 16th-century Vicentine noble family and its relationship to a Palladian villa.

In July 1578 Orazio Godi, one of Pietro's seven sons, while resident in the Villa at Lonedo, murdered his neighbor Tomasso Piovene, another Vicentine patrician whose villa and lands lay directly adjacent to those of the Godi. The exact reasons for the crime are unknown but almost certainly relate to a series of disputes over mill rights on the river Astico and rivalry over land acquisitions that go back to the early 1530s, becoming acute over the course of the 1570s (BCB 1578a). The result was that the Venetian Council of ten sentenced Orazio to perpetual banishment "from Venice and its district, and all other lands and places under its dominion, whether by sea or by land" (BCB 1578b). They also required that Orazio's house in Vicenza be leveled and that his estates be confiscated to the Piovene family. Since the Godi lands were administered as a unit between the seven brothers this required that they declare the part belonging to the exiled Orazio. Their refusal to do so and their probable destruction of any documents relating to the division accounts for our ignorance of their arrangements. However, within months the Venetian state sent out a surveyor, one Giovan Battista Remi, an engineer-surveyor or *perito*, from Venice, to make an inventory of the entire landed possessions of the Godi family. His survey (BCB 1578c), which consisted of a description of each land parcel, its size, use, and value has survived in the family archive, as has a related pair of maps revealing the shape and location of each piece of land, accurately surveyed (BCB 1578c, 1580) (Fig. 3). These maps divide the parcels into seven parts and allocate one-seventh to Orazio for transfer to his victim's family. The maps and inventory, works of remarkable quality and some artistry, particularly given the difficulties under which they was produced, allow us to place the Godi villa in the full context of its economic and social landscape.

According to Remi's estimation the Godi brothers owned and controled land in two distinct regions of the province of Vicenza: at Lonedo di Lugo in the valley of the river Astico and the Piedmont hills around the town of Breganze; and in the floodplain of the Bacchiglione river at Grisignano south-east of Vicenza on the route to Padua. The two areas represent very different physical environments. The first is among the foothills rising towards the Alps, a tapestry of meadow and arable land, planted with vines and fruit or nut trees. The fields were divided for the most part into small parcels with an intensive embroidery of trees and vines over the slopes of green meadow. Here is the site chosen by Giacomo for his villa. However it lies above the line of *fontanili*, natural springs in the limestone, which stretch south of the Alps across the north of the Italian plain, and feed the slow-flowing rivers that empty into the Venetian lagoon. Consequently, for much of the year the lands are dry and their potential can only be fully realized with irrigation. The second area of Godi lands lay in the ill-drained plain where spring floodwaters from the Alps spread easily over the fields as rivers are constricted on the flat lands by the Berici and Euganean hills rising abruptly in the heart of the Veneto. Here the need was for embankment and drainage, a process actively encouraged by the Venetian Republic during the 16th century. The two regions of the province thus have one aspect of their environment in common: both are hydraulic landscapes, and hydraulic landscapes because of the costs of their construction and maintenance and of the organization necessary to their operation, generally benefit those with political and economic power.

According to the Remi survey, whose accuracy is far more dependable than the submissions made for tax estimates by the Godi themselves (BCB 1541), the family possessed a total of 1,823 Vicentine *campi*, approximately 705 ha, in the province, and had made loans secured on a large number of other land parcels which paid a regular income in interest. In the Lugo area a total of 705 *campi* was divided into 106 parcels with an average area of 6.65 *campi*. This average means very little as 75% of the parcels were less than 5 *campi* in size, while the largest block, that surrounding the villa itself was 129 *campi* in area (the size distribution is given in Table 1). In the Grisignano area, by contrast, blocks were larger: 42 parcels occupied 1,118 *campi* (432.06 ha) giving an average parcel size of 6.64 *campi* and 8 parcels accounted for 70% of the land. In both areas the vast majority of land parcels are recorded under the general land use heading of *arativi-piantate* or *prativi-piantate*, that is arable or pasture land strung with vines supported on live support trees, but in the northern estate 12.5% of the land is in orchard (*brolo*). The value of this land is difficult to estimate since land quality was highly variable, and the existence of irrigation channels or drainage ditches could radically increase it. An indication of land value may be given by the estimate made for an arbitration in Montegalda (near the Grisignano estate) in 1572 (BCB 1572) which valued lands at between 15 and 60 ducats per *campo* with an average of 38.5 ducats per *campo*. Using the same figures the value of the Godi estates at Lugo and Grisignano would have been in the order of 70,000 ducats, a considerable fortune.

Table 7.1   Godi estates in 1578 (Remi survey).

		Lugo	Grisignano	Total
Area	(Campi)	705	1118	1823
	(Hectares)	272	432	704
Parcels		106	42	148
Ave. parcel	(Campi)	6.65	26.64	
Size	(Hectares)	2.57	10.29	
Size distribution of parcels				
	<1	30	4	
(Campi)	1–5	49	11	
	6–10	9	5	
	11–25	7	10	
	>25	8	12	

	Lugo		Grisignano	
	Parcels	Campi	Parcels	Campi
Land use				
Arable with vine	78	490.5	28	569
Pasture with vine	20	151.75	7	237.75
Arable and pasture	2	14	2	232
Orchard/garden	5	34	4	15
Woodland/pasture	1	12	–	–
Not specified	–	–	1	29

Where land had been consolidated into large blocks it was worth the capital expenditure of irrigation or drainage. Such land could then be used for crops yielding a high commercial return. Specialized vines, orchard fruits, and vegetables for the use of the villa were the favored uses of irrigated land (Ciriaconi 1981), while the most profitable crop on the reclaimed lower lands was rice (Stella 1956, Ciriaconi 1981 pp. 138–50). A certain amount of land drainage and irrigation had been practiced in the Veneto for many years but from the mid-16th century the pace and scale of operations increased dramatically (Mazzi 1927). For the same reason that urban patrician families who had formerly invested in trade were now amassing land, they were also improving it. Venice itself, among the five largest European cities, presented a huge market for grains and agricultural foodstuffs. Traditionally the city had supplied this demand by sea from southern Italy and islands under its dominion in the eastern mediterranean: Crete and Rhodes. With the loss of some of these traditional granaries and the threat to the sea lanes from the Turks, the Republic turned increasingly to its landed territories to supply its food needs. A policy of land development and improvement (Borelli 1979, Beltrami 1961) was formally instituted during the 1540s involving great schemes of drainage in the lower valleys of the Po, Adige, Brenta, Bacchiglione and Piave rivers, and a formal system of water administration and approval of irrigation schemes under the ministries for Water and Uncultivated Lands. Such a co-ordinated policy called upon the skills of water engineers and surveyors, the *periti*, of whom Remi was one. These men, whose skills and training often overlapped with those of architects, practiced the rapidly developing techniques of leveling, land measurement and cartography (Mazzi 1927). Their knowledge was incorporated in the survey manuals published in considerable numbers in Venice at this time, and this too depended fundamentally on an understanding of the principles of Euclidian geometry. The landscapes they devised, on paper in their maps and *disegni* and on the ground in the forms of new fields, drainage and irrigation channels, regulated rivers and aqueducts were mathematical: the imposition of geometry over nature.

That the Godi were fully involved in this process of land improvement and commercial farming is indicated in the record of their applications to the *Magistratura ai Beni Inculti* for permission to divert waters and irrigate lands around the villa at Lugo and their participation in *consorti* for the drainage of lands in the southeastern Vicentino. At Lugo they not only applied for construction of aqueducts to irrigate the orchard lands behind the villa, part of the large consolidated block of land shown on the 1578 map, but for joint irrigation and mill-races on the rivers Chiavon and Astico in the valley below the villa (ASV 1562, 1575, 1594). Members of the Godi family are recorded as *presidenti* of *consorzi* for large-scale drainage schemes in the southern Vicentino in 1592 and 1599 (ASV 1592, 1599), and in 1594 Màrc'Antonio Godi made application for the flooding of a 100 *campi* rice field in Sossano, an application granted literally one month before the Venetian State passed legislation to prevent any more land being converted to rice because of the damaging effects of this highly profitable but land and labor demanding crop on the peasant economy of the Veneto (Ciriacono 1981).

The Villa Godi was thus integrated into the hydraulic landscape which was being systematically created in the Venetian territories, a modernizing landscape, for part of which the villa acted as the administrative center, a place of work, and the heart of a sophisticated agrarian economic system. When Orazio or Pietro Godi looked through the windows of their villa, their backs to the representations of classical landscapes, they could observe surveyed and newly rationalized field patterns, mills, and water-wheels along the rivers and the improved irrigated lands which promised abundant crops of wheat, vines, vegetables, nuts, and fruits, tended by a subordinate peasantry bound to the family by debt and the annual homage payments that debt entailed.

The cultural world of the Godi

Throughout my description of the Godi villa, its architecture, decoration, and the landscapes, real and illusory, in which it was located, mathematics and Euclidian geometry have been a recurrent motif. They underlay many of the techniques employed in the construction and representation of this place. But in the cultural world of late 16th-century Vicentine patricians mathematics and geometry played a far broader role than their mere employment as techniques. Certainly they were essential to the *mechanical arts*: of engineering, construction survey, cartography, and the like. But they also formed the basis of the *liberal arts*: music, painting, theater, and so on. And, as we have observed, they provided the foundation of architecture, that art which mediated the liberal and mechanical, which allowed people to reproduce in the temples they built that greater temple constructed by the supreme artificer, God Himself, architect of the great machine of the world (Barbieri 1983). It is this cultural world, in which the Godi and their architect, Palladio, participated, that I want to examine now, for it informs the entire conception of the villa and its representations and is as integral to the relationship between place and power as the economic and social world of the Godi.

Among the 68 names listed as members of the Olympic Academy of Vicenza in 1569 we find Horatio Godi, the latinized version of the Venetian Orazio. The Olympic Academy was founded in 1555 by a small group of Vicentines, nobles, and educated professional men, to act as a forum for humanist discourse (Ziggioti 1555). It was modeled upon the famous Florentine Academy of the late 15th century whose members had included such luminaries as Marsilio Ficino and Pico della Mirandola. The Olympic Academy, many of whose members' youth had been shaped by Giangiorgio Trissino, was but one of many such academies in provincial Italian cities at this period. At its inception it welcomed educated men of non-noble birth like Andrea Palladio, but very soon it became an exclusively aristocratic forum, reflecting the increasing tendency towards a closure of access to humanist culture and power generally around the nobility of the Venetian cities (Ventura 1964). Its original statute declared its aims and philosophy:

> The Olympic academicians are united in a single spirit and wish: whence it is not to be marvelled at, for all roads lead to a single end, that each one of them desires to improve all the sciences, and particularly that of Mathematics; which is the true ornament of all those who have a noble and virtuous soul. (Ziggioti 1555 p. 103)

The Academy had its own library which included the major classical works of literature and philosophy, Euclid's *Elements*, Renaissance classics like Alberti's *Architecture* and Vasari's *Lives*, Dante, and books on geography, astronomy, and alchemy, as well as sculptures by Sansovino and Michelangelo, astrolabes, geographical maps, and almanacs. It sponsored plays and lectures and is best remembered today for the Olympic Theatre, designed by Palladio as a reconstruction of a Classical Greek theater with a permanent stage set by Vincenzo Scamozzi representing the townscape of an ideal city, the appropriate context for the performance of tragedy.

The Olympic Academy was the focus of "high" culture in late 16th-century Vicenza. Its interests betray the acceptance among the Vicentine elite of a very particular cosmology, or world view, characteristic of the late Renaissance in Europe. It was a cosmology that would be swept away in the religious, economic, and intellectual turmoil of the early years of the next century: the years of persecution, witch-craze, depression, and Cartesian revolution, and is thus largely unfamiliar to us (Yates 1979). Without going into the complexities of this very sophisticated and arcane culture it is difficult to do it justice to a modern audience. But at the risk of trivializing we may say that the intellectual culture of the Vicentine academicians was founded on a holistic view of the universe as governed by what the 15th-century humanist Leon Battista Alberti had called a "secret discourse" (Alberti 1965), a pattern of symmetries and proportions that ran through the structure of creation producing harmony, from the macrocosm of the universe itself with its revolving spheres, to the earth with its four natural elements, to the body of man, the microcosm (Yates 1979, Barbieri 1983). The key to understanding this structure, given to man alone through the gift of reason, lay in mathematics, and particularly those branches of mathematics that dealt with proportion and geometry. In the study of mathematics man could come to know God's plan and emulate it in his own "arts", both mechanical and liberal, most perfectly in architecture.

Among the 21 founder members of the Olympic Academy is listed Silvio Belli, an engineer-surveyor and general mathematical practitioner. In 1575 his book, *Della Proportione e Proportionalità* (1573), formed the subject of one of the Academy's meetings, and he lectured to its members more than once. *Della Proportione . . .* opens with a dedication in which Belli makes reference to Plato's respect for mathematics: "the mathematical disciplines entice, engergise, power and direct reason, intelligence and contemplation in the direction of truth" (Belli 1573). The book has two parts, the first devoted to the principles of proportion and ratio in numbers and quantities, developing the argument that proportion is fundamental to the natural order whose governing principles can be disclosed through mathematics. The second part deals with *proportionalità* and is altogether more erudite. Essentially it is concerned with the relationship between numbers, geometry, and musical harmonics. The interest in this relationship goes back to Pythagoras and was popularized in the early Renaissance in the writings of 15th-century Florentine humanists (Wittkower 1962, Yates 1964). Its significance derived from the fact that the same mathematical and proportional relationships to be found in matter (e.g., proportionate lengths of a lute string) could also be discovered in non-material phenomena (e.g., the scale of sounds emitted by the string if plucked at those different lengths). From this it was possible to construct infinitely elaborate number systems and relationships which could then be traced throughout the natural world and beyond. Astronomical and astrological readings, interpretations, and predictions were all founded in number; time was incorporated through the calculation of calendars and their numerical proportions; and a whole scientific and philosophical corpus was erected. The plan of the Villa Godi is based upon such a set of proportional numbers. In the hands of some devotees the incorporation of Cabala numerology and related arcana led this philosophy towards the occult, but there is no strong suggestion of the Olympic Academy having particular leanings in that direction, other than the reference in its library collection to alchemical texts.

The practice of magic based upon the mathematical philosophy was for some the ultimate power, a power that extended over the natural, celestial, and supercelestial worlds (Yates 1979, pp. 37–47). But it is not necessary to go these lengths to realize that the culture of the Olympic academicians was profoundly aligned to power, both political power and power over the natural environment. Mathematics and the pursuit of reason empowered the individual. It was regarded as the source of *virtù*, the mark of the noblest soul. In Italy during the second part of the 16th century nobility was increasingly being tied to birth as the elites of cities like Vicenza closed off traditional access to their ranks by "new men", men who had made fortunes in trade or manufacture (Ventura 1964). In 1567 the Noble Council of Vicenza, whose seats were occupied by the same group of men who styled themselves Olympic Academicians, declared

> that no one may be elected to the College of Magnificent Deputies, to the Council of One Hundred, to the Fisc, to the Consulate, to the Vicariate, who has not possessed citizenship for at least one hundred years before that election, or who has practised, either himself, or his father, any mechanical art. . . . (Ventura 1964, p. 363)

In other words, the oligarchy who ruled the city and its province specified a strict relationship between the liberal arts and political authority. It was a closed logic whereby *virtù* defined fitness to rule, and the attainment of *virtù* was possible only for those born into the ruling elite. Participation in the culture of the mathematical philosophy was thus a mark of political authority. The names of professional men like Palladio, who had originally been accepted into the Olympic Academy were to disappear from its membership lists by the 1570s. Their understanding of the mathematical philosophy allowed them to act as practitioners, spokesmen for the culture of the ruling class, but they were denied any political role.

Such practitioners did however deploy the principles of the mathematical philosophy in the exercise of control over the natural world. They were the engineers, the surveyors, the architects, patronized by the nobility. An earlier work by Silvio Belli, *Libro di Misurar con la Vista* (1565) is a practical manual of survey, describing different methods for determining heights and depths, distance and area. It was but one of a large number of such texts published in Venice and its provinces in these years, and Belli himself was employed by the State to conduct a survey of communal boundaries (Schulz 1976). A similar text, published in Venice and written by Nicolo Tartaglia, first translator of Euclid's *Elements* into Italian, was titled *La Nuova Scientia* (1550). The "new science" was that of ballistics in which geometry was employed to determine the trajectories of cannonfire in order that the increasingly powerful and sophisticated artillery pieces of the period could be used effectively. In military science geometry was also applied to determining infantry formations and most dramatically, as far as the contemporary landscape was concerned, to the design and construction of fortifications. In the aftermath of the Cambrai wars when Venice had come close to forfeiting its entire landed patrimony in Italy, the Republic initiated the reconstruction of defensive walls around the provincial cities (Hale 1968). The results of this work, designed by the engineers as exercises in applied geometry, are still visible in the great bastions and angled revets of Verona or Padua. The apotheosis of this work was the fortress town of Palmanova, designed by Palladio's successor Vincenzo Scamozzi in the borderlands between Venice and the Austrian Empire, a town whose ground plan is that of an ideal city, that utopian ideal formulated endlessly on paper by late Renaissance architects and embodying entirely the social vision of the ruling class (Cosgrove 1982). It comes as no surprise that the same architect designed the permanent stage set of Palladio's Olympic Theatre in Vicenza, a series of perspectives down the streets of a classical city where the dramatic performances of the Olympic academicians were staged.

Geometry was thus essential to warfare, the crudest exercise of power. But as we have already observed it was equally essential to transforming the environment of the *terraferma*. The cartographers, the surveyors, the hydraulic engineers, frequently of course the same men, all used its principles in their labors. They wrote mathematical texts and employed such handbooks for their work (Cosgrove 1988). And this whole corpus of practical knowledge was theoretically integrated in the treatises written by the architects: in Palladio's *Four Books of Architecture* (1570) and Scamozzi's *L'Idea dell'Architettura Universale* (1615). Architecture mediated the mechanical arts of the engineers and the liberal arts of the humanists and nobility. It was the architects who designed the Council chambers, the palaces, the villas, and the churches of the nobility, who provided the physical environments in which those who exercised political power lived, worked, and prayed. Their buildings articulated the world picture of the ruling class. Thus Palladio, architect of the Villa Godi, designed palaces and villas for most of the great family clans of Vicenza in the years between 1535 and 1580, and of course in his later years he went on to become the successor to Jacopo Sansovino as architect of the most important public buildings of Venice itself, his three great churches transforming the water spaces that front the Doge's Palace.

Politics and place

If we return now to the frescoes of the Villa Godi in the light of the context I have sketched out they take on a more complex significance than merely the playful decoration of a summer retreat for an aristocratic family. They may now be read as a pictorial "placing" of the Godi within a representation of their world. The gentlemen who draw aside the velvet hangings to reveal the scene, or whose children sit on the illusory ledge overlooking the painted landscape, are quite properly garbed in the costume of the mid-16th century, that of the inhabitants of the villa. The landscape they gaze upon is a refraction of the landscape surrounding the villa both in the real sense that it is recognizable topographically and in some of its vernacular features, and the imaginary sense that its Classical buildings represent a past utopian world in which the nobility participated in the Olympic Academy, in the public and private palaces of the city of Vicenza and in the literary and philosophical discourses they held in the villa. This participation was more than a recovery of the past. In the eyes of the Godi their own world shared with that of Ancient Greece and Rome an understanding of universal laws which governed creation, an order and harmony in nature and society. These were reassuring images, illusions created by the laws of perspective which could be held in the mind as one turned outwards towards the subject hills and fields and farms of the family estate, themselves ordered by the same rules of proportion and harmony displayed in the frescoes, the surveyed and improved lands making a setting for the proportioned dignity of the villa. The power to create and appropriate an image of landscape and nature on the walls of the villa not only illustrates the power to create and appropriate landscape and nature in reality, but both express a single authority, at once cultural and economic, aesthetic and practical, believed to derive from nature itself.

In the Godi landscapes, as so often in the history of landscape representation, the lived world of power and authority is re-presented in an image of place. The flux and uncertainty of daily life are transformed into an immutable image of harmony. The frescoes narrate a world which is neither real in the sense that it existed empirically on the Godi estates, nor false in the sense that it demonstrably did not, at least in the experience of the Godi. Rather we should view these paintings, the villa itself and the landscape in which it was set as equal participants in the cultural process of creating human meaning and articulating power within the social and environmental relations of the Venetian *terraferma* in the mid-16th century.

References

Ackerman, J. 1966. *Palladio*. Harmondsworth: Penguin.
Alberti, L. B. 1965. *Ten books on architecture*. Trans. J. Leoni, 1755. Facs. London: Alec Tiranti.
Archivio di Stato di Venezia (ASV), Provv.Beni Inculti (PBI) Proci. B41, reg. 2 Godi (6.2.1562, 3.2.1575, 11.5.1575, 28.7.1575, 16.8.1594).
ASV, PBI Consorzi B807, reg. 3 (15.4.1592), reg. 5 (7.3.1599).
Barbieri, F. 1970. *The Basilica of Andrea Palladio*. Corpus Palladianum, Vol. 2. London: Pennsylvania State University Press.
Barbieri, F. 1983. *Andrea Palladio e la cultura Veneta del Rinascimento*. Roma: Il Veltro.
Barrell, J. 1980. *The dark side of the landscape: the rural poor in English painting 1730–1840*. Cambridge: Cambridge University Press.
Battilotti, D. 1980. *Vicenza al Tempo di Andrea Palladio attraverso i Libri dell' Estimo del 1563–1564*. Vicenza: Accademia Olimpico.

BCB, *Arch. Godi*, Proco. 13.154.13 (1541).
BCB, *Arch. Godi*, Proco. 1.16, 64r–76v (22.3.1572).
BCB, Vicenza, *Archivio Godi*, Catastico, MXCII 4405, 1536.
BCB, *Arch. Godi*, Proco. 139.2801, 7v (7.4.1573).
BCB, *Arch. Godi*, Proco. 119.2306 (30.7.1578-20.11.1578).a
BCB, *Arch. Godi*, Proco. 119.2306, 21v (17.10.1578).b
BCB, *Arch. Godi*, Proco. 1.16, 26r-64v (1578c); 76r-104v (23.3.1580).
These maps are held separately at the BCB and numbered 2-XVI Lonedo and
    4-XVI Grisignano respectively.
Belli, S. 1573. *Della proportione e proportionalità*. Venezia.
Beltrami, D. 1961. *Penetrazione economica dei Veneziani in terraferma: forze di lavoro
    e proprietà fondiaria nelle campagne Venete dei Secoli XVII e XVIII: Venezia. Instituto
    per la Collaborozione Sociale.*
Berger, J. 1976. *A fortunate man.* London: Writers & Readers.
Borelli, G. (ed.) 1979. *Uomini e acqua nella republica Veneta tra Secolo XVI e Secolo
    XVIII: il tratto Veronese dell'adige.* Verona: Borelhi.
Cevese, R. 1970. *Ville della Provincia di Vicenza.* 2 vv. Milano: SISAR, pp. 81–97.
Ciriacono, S. 1981. Investimenti capitalistici e coltura irrigue. La coniuntura agricola
    nella terraferma veneta (secoli XVI e XVII). In *Venezia e la terraferma*, Tagliaferri
    (ed.) Milano: Guiffre, pp. 123–58.
Corazzol, G. 1979. *Fitti e Livelli a Grano. Un aspetto del credito nel Veneto del '500.*
    Milano: Angelo.
Cosgrove, D. 1982. Problems of interpreting the symbolism of past landscapes.
    In *Period and place. Research methods in historical geography*, A. R. H. Baker & M.
    Billings (eds), pp. 220–32. Cambridge: Cambridge University Press.
Cosgrove, D. 1984. *Social formation and symbolic landscape.* London: Croom Helm.
Cosgrove, D. 1985. Prospect, perspective and the evolution of the landscape idea.
    Transactions, Inst. Br. Geogrs **10**, (1) pp. 41–62.
Cosgrove, D. 1988. The geometry of landscape: practical and speculative arts
    in the sixteenth-century Venetian land territories. In *The iconography of land-
    scape*, D. Cosgrove & S. Daniels (eds). Cambridge: Cambridge University
    Press.
Cosgrove, D. & S. Daniels 1988. "Iconography and landscape" in *The
    iconography of landscape.* Cambridge: Cambridge University Press.
Crescentio, P. B. 1561 *De Agricoltura Libre XII.* Sansovino.
Duncan, J. 1985. Individual action and political power: a structuration perspective. In
    The future of geography, R. J. Johnston (ed.), pp. 174–89. London: Methuen.
Eyles, J. 1985. *Senses of place.* Warrington: Silverbrook.
Gallo, A. 1565. *Dieci Giornate della Vera Agricoltura.* Vinegia: Farmi.
Gold, J. R. 1985. From "metropolis" to "the city". Film versions of the future city
    1919–1939. In *Geography, the media and popular culture*, J. Gold & J. Burgess
    (eds). London: Croom Helm.
Hale, J. 1968. Francesco Tensini and the fortification of Vicenza. *Studi Veneziani*
    10, pp. 202–45.
Harley, J. B. 1988. Maps, knowledge and power. In *The iconography of landscape*,
    Cosgrove & Daniels (eds). Cambridge: Cambridge University Press.
Mantese, G. 1966–67. Tre capelle gentilizie nelle chiese di S. Lorenzo, S. Michele
    e S Corona di Vicenza. *Odeo Olimpico* VII, pp. 227-58, n.3.
Mazzi, B. nd. *La Villa Godi di Lonedo.* Milano: Orga, 12.
Mozzi, U. 1927. *I magistrati Venete alle acque e alle bonifiche.* Bologna: Nicola
    Zanichelli.
Nicodemi, G. 1967. *La Pittura Italiana dell'800 a Villa Godi Valmarana.* Milano: Orga.

Novak, B. 1980. *Nature and culture: American landscape and painting 1825–1875*. London: Thames & Hudson.

Palladio, A. 1965. *The four books of architecture*. Trans. Isaac Ware, 1738. New York: Dover, 52.

Pigafetta, F. 1974. *La Descrizione del territorio e del Contado di Vicenza (1602–1603)*. A. da Schio & F. Barbieri (eds). Vicenza: Neri Pozza, 46.

Puppi, L. 1972. The villa garden in the Veneto. In *The Italian garden*, D. R. Coffin (ed.), pp. 83–114. Washington, DC: Dumbarton Oaks.

Rossi, V. 1982. *Villa Godi-Malinverni*. Schio: Pasqualotto.

Rumor, S. 1899. *Il Blasone Vicentino descritto e storicamente illustrato con cento ventiquattro Stemmi incise e colorati*. Venezia:. Visenti.

Relph, E. 1976. *Place and placelessness*. London: Pion.

Samuels, M. 1979. The biography of landscape. In *The interpretation of ordinary landscapes*, D. Meinig (ed.), pp. 55–88. Oxford: Oxford University Press.

Sartori, P. L. 1981. Gli scrittori veneti d'agraria del cinquecento e del primo seicento. Tra realta e utopia. In *Atti del Convegno: Venezia e la Terraferma attraverso le relazione dei rettori*, A. Tagliaferri (ed.), pp. 261–310. Milano: Giuffre.

Schulz, J. 1976. New maps and landscape drawings by Cristoforo Sorte. *Mitteilungen der Kunsthistorischen Institutes in Florenz* **XX**, (1), pp. 107–26.

Solkin, D. 1982. *The landscape of reaction. The paintings of Richard Wilson*. London: Tate Gallery.

Stella, A. 1956. La crisi economica veneziana della seconda meta del secolo XVI. *Archivio Veneto* **XVI**, (58), pp. 17–69.

Ventura, A. 1964. *Nobilità e Popolo nella società Veneta del '400 e '500*. Bari: Laterza.

Ventura, A. 1968. Considerazione sull'agricoltura veneta e sulla accumulazione originaria del capitale nei secoli XVI e XVII. *Studi Storici* **IX**, (3–4) pp. 674–722.

Wittkower, R. 1962. *Architectural principles in the age of humanism*. London: Alec Tiranti.

Yates, F. 1979. *The occult philosophy in the Elizabethan Age*. London: Routledge & Kegan Paul.

Yates, F. 1964. *Giordano Bruno and the hermetic tradition in the Renaissance*. London: Routledge & Kegan Paul.

BCB Ziggioti (Gonz. 21.11.2) "Statui e Membri dell'Accademia Olimpico', 1555.

8

Place, meaning, and discourse in French language geography

VINCENT BERDOULAY

The notion of place, whether the word is actually used or not, is of increasingly widespread interest. In this chapter I will review contributions made by French language geography to the study of place.

I will be commenting here on geographical research produced by scholars who are from French-speaking areas or who have chosen to use French as a medium of communication, for although French geographers have produced a major portion of this work, one must also take into account a growing body of literature which is not produced in France and yet contributes to the structuring of a specific intellectual space. This is increasingly realized by non-Francophone observers (for instance, Copeta 1986). In the same spirit, I will not strictly limit myself to the work of researchers who are institutionally labeled geographers, but will extend my comments to the work of individuals who, broadly speaking, have contributed to geographic thought. In this regard, the work of a number of French historians who have had a long-standing interest in geographical issues, as exemplified by the works of the *Annales* school, comes to mind. My aim, however, is not to do a full literature review, but rather to mention contributions which I find helpful to characterize a few place-related themes.

Why has there been an interest in place (what the French term *lieu*)? Is the study of places different from regional geography? My view is that we should differentiate the notion of place from other terms in order to make it useful. The term place (*lieu*) is used among French language geographers in an informal sense. As such it is generally not used as a research-inducing concept, as is often the case in Anglo-American geography. Among Francophones it usually refers to a portion of geographical space, with the idea of location ascribed to it. Such an umbrella usage encompasses everything from "*la theorie des lieux centraux*" (central place theory) to the often-quoted remark from Vidal de la Blache "*la géographie est la science des lieux et non celle des hommes*" (geography is the science of places, not that of men). There are, however, two overlapping types of connotations which it often has.

First of all, the notion of place connotes a concern for integration, for wholes. In other words, it implies what one might term a humanistic view.

Interestingly enough, the term *lieu* is used in the Aristotelian tradition to designate what is considered to be the proper location of all things and, consequently, what induces their movement toward it. It is thus tied to formal and final causes. This leads to the second connotation that the term often carries. The idea of place implies a meaningful portion of geographical space. Place involves meaning for the people who build it, or live in it, or visit it, or study it. This is why it is close to notions such as territoriality (*territorialité*) and landscape (*paysage*), in the sense that there is a special, usually emotional, link between people and place – the latter being understood as a concrete and very specific area.

Again, a lexical remark may be enlightening. The French word *sens* stands for meaning, but it may also designate direction, way. This semantic association points to the interdependence between meaning and movement as astutely noted by Greimas (1976) and suggested by the Aristotelian usage of the term. We are now concerned with meaning as well as with direction, movement, or change when we think of place. The latter can be viewed as a particular application of the idea of region. This makes the study of place more focused than regional geography as a whole. It is part, however, of the same genre of scientific discourse.

The major themes in this chapter – place, meaning, and discourse – have thus been suggested. I will now attempt to shed some light on the interconnections between them. Since French geography owes much to the foundational work of Vidal de la Blache, I will first summarize the Vidalian legacy as far as it relates to place. I will then look at more recent research on the meaning of place as revealed by its formal structure. However, I will also examine works which focus upon the semantics or pragmatics of place and, consequently, cut across the related notion of landscape. I will then discuss the problem of the emergence of new forms and meanings which is raised in the previous discussion. Finally, I will raise another issue: the necessity for geographers to take into account the discursive level which exists both in the functioning of place as well as in our analyses. I will thus try to leave the reader with a better view of how place and geographic discourse may be two sides of the same reality in a process of emergence.

The Vidalian legacy

Undoubtedly, the foundation of geography laid by Vidal de la Blache has a lot in common with the concern for place. But often, his legacy has been grossly misunderstood. For instance, the English language literature has long been plagued with a persistent tradition of stereotyping the Vidalian contribution. It is often equated with (a) a radical possibilism (man is free to do what he wishes in the face of a passive nature) and (b) the regional monographic approach (all geography consists of well-written descriptions of unique regions). Fortunately, there are encouraging signs of a renewed interest in the actual contribution of the early Vidalian school of thought, and particularly of Vidal himself. This is only fitting as he has a comparable significance to geography as Emile Durkheim or Max Weber have to present day sociology. We are even now witnessing the coming of age of a true Vidaliana! (e.g., Claval & Nardy 1968, Pinchemel 1970–72, Grau 1977, Buttimer 1978, Berdoulay 1981, 1983, Gregory 1981, Garcia Ballesteros 1983, Andrews 1984, Nozawa 1984).

As I have tried to show in a previous book (Berdoulay 1981), Vidalian possibilism is best characterized as a broad theoretical framework devised for avoiding both radical possibilism and environmental determinism. The point was to focus not on one-directional deterministic influences but rather on the interaction of all the elements pertinent to the understanding of man–nature relationships. In this respect, we cannot confine this approach to the study of so called "vertical", i.e., local ecological relationships alone, for the possibilistic perspective entails spatial relationships as well – be they physical or human. This was well noted by Lukermann (1965), who observed the role of the "circulation" and "ensemble" themes in French geography. Vidal's book *La France de l'Est* (1917) is a remarkable example of the importance he attributed to the "*vie de relation*" (different from the "*vie locale*") in shaping that region. This is all the more noteworthy in that he was well aware of the importance to be given to physiographic features for explaining the regional organization of France (Vidal, 1888-89).

In order to dispel the stereotypes regarding Vidalian geography, one must realize that regional studies were one of the means toward the geographer's goal of understanding man-nature relationships. It is also essential to note that the Vidalian regional approach was applied at various scales. For this reason, Vidal's *Tableau de la Géographie de la France* (1903) is quite different from *La France de l'Est* (1917), or Demangeon's *La Plaine Picarde* (1905) from *L'Empire Britannique* (1923). Also noteworthy is the minute attention brought to very small areas such as the local *pays* of France (Gallois 1908, Brunhes 1910 on the Val d'Anniviers). At all scales, although it is more apparent at the local level, the feelings of the inhabitants are used to explain the very special links which exist between man and his environment. These feelings have resulted from the long-time and effort-loaded interaction between them. Fittingly enough, this theme nurtured nationalism, but it could also be used to foster a regionalist ideology or, for others, a global consciousness. A specific scale was not a prerequisite for the regional approach.

As we have seen, the Vidalian thrust in geography is compatible with the current interest in place. It was very attentive to the environment as experienced by people. The concern for people's plans, worries, initiatives, and efforts gave this geography the highly humanistic overtones which have frequently been noted by non-French commentators (Buttimer 1971, Ley 1977, Ley & Samuels 1978). However, one must underline here that the Vidalian perspective on place is not to be grounded in an idealist philosophy of man nor in some immanent properties of the environment (such as the *genius loci*). Rather, it is entirely focused on the interactions which permeate the man-nature system.

Furthermore, the Vidalian approach to place has to be viewed within the perspective of emergence. Fundamental to the epistemology of Vidal's school (Berdoulay 1981) is the problem of emergence. New forms of landscapes and corresponding social arrangements keep appearing on the surface of the earth. The organization underlying them achieves, for a period of time, some kind of equilibrium, or stability, which the geographer can precisely analyze, whatever the degree of contingency which sparked the reorganization process. How a place came to be and what makes it new are essential underlying questions in this perspective. Clearly, Vidalian interpretations of place have varied according to the author and the time. A good example is provided by the Mediterranean countries, successively studied by Vidal himself, as well as Jacques Weurlesse, Pierre Birot, Pierre Deffontaines, etc. (Claval 1986).

Historians and place

With its strong historical dimension, the Vidalian school quickly attracted the attention of historians. They were quick to grasp the significance of both the possibilist perspective and of regional studies for their discipline. In fact, the so-called *Annales* school, launched by Lucien Febvre, is known for its favorable leanings toward geography (Mann 1971).

Lucien Febvre praised the Vidalian school in *La terre et l'évolution humaine* (1923) for its possibilistic approach. He especially heralded its famous regional studies. But in so doing he helped diffuse the view that geography was an overly empirical and ancillary discipline in relation to history (Lacoste 1982: postface). Whatever one's opinion on this, it is clear that historians who were associated with the *Annales* repeatedly turned toward geography as a source of information and even inspiration. The integrative, comprehensive regional studies by the Vidalian school have served as the mainsprings of regional histories – an early concern started by Henri Berr (1903) and even Lucien Febvre (1905) himself. Rural history and rural geography had profitable mutual exchanges during the inter-war period. This was greatly due to the contributions of such historians as Gaston Roupnel (1932), Marc Bloch (1931), and Roger Dion (1934).

But, with the rise of the hegemony of the *Annales* school, geography progressively came to lose the fascination it had held for historians (Coutau-Bégarie 1983). Under the influence of Ernest Labrousse after World War II, French historians increasingly turned toward the study of statistical and economic "conjunctures". Their works became removed from the concern for place. It is mostly with Fernand Braudel and his much praised *La Méditerranée et le monde méditerranéen à l'époque de Philippe II* (1945) that geography appeared again as an essential component of the historical approach. Here the emphasis was upon a vast territory, and place took on a meso-scale significance, much beyond the individual's daily life. However, this "geographical history" was little imitated, probably because of the massive erudition it required. This geographical history was also cut off because of its scale from the more micro-scale regional studies written by the geographers of the time.

Within the last quarter of a century, regional history has sprung up again at the local and meso-scales, and a limited *rapprochement* with geography has occurred (Coutau-Bégarie 1983, p. 75). E. Le Roy-Ladurie (1966), G. Duby & E. Wallon (1975), and very recently F. Braudel (1986) have been very instrumental in this respect. The periodical *Études rurales* has served as a forum for scholars from the two disciplines. The place perspective, however, is not central to these interdisciplinary encounters. The main emphasis of the *Annales* school today is upon *mentalités*, an ill-defined concept encompassing much of social-psychological and cultural history. For these historians, place just constitutes a setting for human action; it does not induce meaning.

Nevertheless, in spite of the dominant thrust of recent French historians, their school of thought does not preclude what Braudel (1986) has done when he turns his interests toward the multi-tiered spatial fabric of France, from the smallest *pays* to the national territory. In this work, the idea of place comes up more at the local level, although not necessarily so. The physical environment is called into play, for instance as a constraint on the spatial interactions among people. The whole approach to place is set by Braudel within the broad integrative view of the *Annales* school. He is particularly fascinated by the place-oriented plurality of French social life. According to him, this constitutes an essential element of the identity and unity of France: "*le singulier au sommet, le pluriel à la base*" (*ibid*, p. 40). Braudel epitomizes some of the Vidalian concerns associated with place, including the emerging characteristics of new wholes, new entities.

Let us now consider recent contributions by French language geographers in order to see what they have added to the Vidalian foundations.

Structuralist Preliminaries

In order to preserve the structural orientation of classical geographic analysis (Claval & Nardy 1968) while at the same time introducing more concern for the meaning of the arrangements thus identified, French geographers have turned toward semiotics. Roger Brunet (1969, 1972) was an early exponent of this semiologically inspired structuralism. His major contribution with respect to place is his flat refusal to fall into the idiographic-nomothetic dichotomy (Brunet 1980). Place has often been associated with the idiographic approach, especially in Anglo-American literature, where researchers have focused upon the unique characteristics of places and the significance of places for individuals. Brunet first identifies choremes (ichorèmesr), the elementary structures in the organization of geographical space. He argues that their actual number seems limited to between 20 and 30. He claims that they are related to the center-periphery model, paths, gradients, discontinuities, dissymmetric relations, diffusion patterns, etc. They thus have geometric characteristics. They enter various combinations and, consequently, produce specific organizations of particular places. These combinations obey various fundamental rules which one can identify. Thus, specific models are derived from elementary models and are related to more general (regional or universal) models (such as those exemplified in classical location theory). Rather than a dichotomy between the idiographic and nomothetic approach, there are just various levels of combination-analysis.

One can see that Brunet's approach is very structural. In this respect, he is in line with a good deal of French research on geographical space. He may even appear so straightforward in his approach that many of the connotations implied by the notion of place seem to be left aside. However, the door is open to more meaning-oriented research as he views choremes as signs, as the building blocks of a semiology of the organization of space. According to semiology a sign is composed of a signifier and a signified. Brunet identifies the signifier as the identified spatial arrangement which either hides or reveals the signified, a strategy of domination over nature or social groups. Such model building, then, consists of identifying the syntactic rules which are like phrases produced by society in order to achieve some project, goal, or finality. The organization of geographical space – and thus of place – by society constitutes its means of reproduction and development. In setting his structuralist approach in this perspective, Brunet anchors his methodology in the all-pervasive context of social meaning. As such he expresses well another characteristic of French geographical research which tries to remain grounded in the social dimension of its subject matter.

There is a definite reluctance among many French geographers to focus on individuals' behaviors or perceptions apart from their social-geographical context. For instance, when Claude Raffestin (1977, 1980) evidences an interest in place-related issues – via his concern for territoriality – he grounds it in the study of the power relations underlying the structuring of a territory (*territoire*). For Raffestin territoriality is made up of all the relationships experienced by individuals with their environment. But these are all mediated by social relations set within production, exchange, and consumption processes. Territoriality is the lived or experienced inside of power relations. It involves not only what is close to individuals but also exterior or remote forces on which local life depends. For instance, Sicilian territoriality is not only a local affair but also partly a fight for preserving an identity, an autonomy versus Northern Italian power which tries to organize the whole national territory according to its own principles and needs. In other words, place, as connotated by territoriality, may have a conflictual nature.

In a related vein, Paul Claval (1973, 1978) searches for a grammar of social relationships as well as for the interaction between space and types of power in order to explain the structure and emergence of place configurations. Gilles Ritchot (1985) also examines the interplay of forms and functions (signifiers and signifieds) in the context of property relations, as a means to ground geographic analysis in the simultaneous consideration of both spatial and social determinants. Whereas, on the one hand, there is a tendency in French language geography to devote attention to spatial structure or form as a constraint on future social structures (Ferrier 1984), on the other hand, there is a simultaneous search for the social determinants of these spatial structures (e.g., Vant 1981). While some researchers stress the structure of place, others are more attentive to the way in which place is socially produced (cf. Raison 1976, Reynaud 1981, Villeneuve 1981, Frémont *et al.* 1984, Bonnemaison, 1986).

Thus, there is a definite interest in French language geography in the syntactic level of place construction. But the emphasis which is placed on the social determinants of this construction has led researchers to explore what place usually connotes for its inhabitants, that is, to give greater consideration to semantics. Let us turn toward this aspect of place.

The place of meaning

Shortly after World War II, Eric Dardel (1952) raised the issue of sense of place. He was echoing contemporary philosophical concerns for phenomenology and existentialism. Whereas French geographers were not trying to make explicit use of these epistemological/ecological orientations as was to be the case in Anglo-American geography, these philosophies reinforced the humanistic bent of much of French geographical thought. Pierre George's (1967) reflections on the diverse perceptions of time seem to me a case in point. These philosophies fostered the emphasis on some themes rather than a search for new methods. In this respect, the study of cognitive representations found a renewed significance (e.g., Frémont 1974, Gallais 1976, Bailly & Ferrier 1986). The concern with living space (*espace de vie*), which consists of the areas often frequented by an individual or a social group, is complemented by the study of social space (*espace social*), which adds to the latter the social relationships underlying them. Lived space (*espace vécu*), then, includes living space, social space, and the values attached to these areas. Consequently, lived space may refer to a single place, or to several places. It involves not only experience but emotions and feelings as well.

What is especially interesting about this view is that it is not necessarily geared to a nostalgic or utopian yearning for harmony. Again, the above-mentioned concern for the social dimension of geography makes Frémont link the notion of lived space to the different modes of production which regulate the relations of men among themselves and with geographical space. This enables him to appreciate how lived space conditions in a different fashion the life of "social marginals". In other words, place may induce marginality, under certain social relations of production (Frémont 1979). For instance, the often idealized idea of place, where rootedness is supposed to promote happiness, is thought to be characteristic of peasant societies. The marginals, then, are those who live in the countryside but who do not own land. Place has different overtones for them. Other rural societies (based on pastoralism, shifting agriculture, etc.) necessarily cannot have the same degree of rootedness to a single limited area. Their sense of place takes different forms, and the concept of marginality is defined differently as well. In urbanized societies, the social division of work induces a dispersed lived space (associated to work, services, and residence) and various types of spatial segregation. As adaptive forms of marginality occur in large cities, this can lead to a strong sense of place in one's living space or part of it, at the expense of an appreciation for the whole urban area. Consequently, the emerging patchwork of turfs makes everyone feel like a marginal in relation to the others.

Furthermore, place does not have to be thought of in terms of area. It may be more significant to view it as a network. For instance, one tends to associate centrality (of urban functions), town center, urbanism, and place. The contemporary gentrification of central neighborhoods and the correlated increase in public life causes some observers to herald the recovery of place – a place where the meaning of urbanism is epitomized. But this apparent revival of urban centrality in a single place may not become universal.

Rather, there is a growing dissociation of the spatial, social, and economic dimensions of centrality (Burgel 1986). The city is still characterized by its potential to maximize social relations, but this is realized in various locations within its limits. The urban place *par excellence* still exists, but it is no more at the center of the city, rather it takes the form of a network of various locations (friends, restaurants, theaters, offices). We see again that the meaning of place is linked to spatial structure, but only in so far as its social determinants are taken into account. In my opinion, this is where the notion of landscape becomes essential.

We now know that the landscape is coded by society. Usually, several codes coexist, as they are linked to different spheres of life, be they social, political, cultural, or economic. However, the same code may pervade all these spheres, as in Japan (Berque 1982). Meaning can then be read in the landscape. When the code is not as pervasive or as well engraved in all landscapes, this type of analysis has limitations. This is why I think that the study of place should benefit from a more operational concept of landscape. In such an approach landscape is viewed as an autonomous level of creation of meaning. While ultimately societal processes are responsible for its production, nevertheless, meaning in the landscape (and thus in place) comes from its own organization. In this perspective, we must take into account physical and organic processes as well. What counts, in fact, is the spatial level of interaction and concatenation of all these various processes, which produce the landscape.

Taking partial inspiration from Soviet and German works, French research has advanced the geographical-ecological study of the landscape (Bertrand 1968, Beroutchachvili & Mathieu 1977, Brossard & Wieber 1980, Phipps 1985, Phipps & Berdoulay 1985). Broadly speaking, the view is that the landscape constitutes a significant level of analysis, since it is embedded in a physical-geographic logic which intersects traditional specialties (geomorphology, climatology, and biogeography, which have their own ends and techniques). The same thing can be proposed for the human side of the landscape. Beside the usual human geographical specialties (cultural, economic, political) and intersecting with them, there is a human geography of the landscape and, in turn, a confrontation between these two autonomous (physical and human) levels. It is indeed possible to note definite convergences between ecological and semiological analyses of the landscape (Berdoulay 1985b). Place, then, can take on a more powerful

131

scientific and conceptual content. It is like the meaning side of the formal concept of landscape.

Now the problem is to understand how such autonomous levels come to be, that is, how space becomes place.

The emergence of place

A conceptualization of the emergence of new entities from chance phenomena already existed at the end of the 19th and beginning of the 20th century (Berdoulay 1981: pp. 201-227). In the mid-19th century, Cournot had characterized chance as an objective reality, as opposed to the Laplacian conception that it was only an expression of our limitations to fully understand the mechanics of the world. Taken up by Emile Boutroux from the 1870s on, the concept of chance was shown to be at work in reality, especially in the subject matter of the life sciences and even more so of the human sciences. He then tried to reconcile science with free will, since the latter appears to be the highest form of contingence. Other contemporaries, such as Poincaré, focused on the significance of the calculus of probabilities as it allowed an approximate definition of a phenomenon (which does not entirely repeat according to our observations), and thus the conventional introduction of a deterministic approach to its study. Such a methodological determinism did not imply necessity and did not preclude free will. These reflections had echoes in the human sciences, especially in history, and in Vidalian geography. The idea was to avoid a mechanistic determinism by allowing the possible effects of contingence in their explanations.

Following Cournot, they recognized that a specific event could result from the chance intersection of independent processes. These results constituted relatively stable phenomena, such as the *pays*, regions, or *genres de vie* that the geographer investigates. This constitutes a scientific approach to the emergence of new organizations (and thus of places) on the surface of the earth. Chance is recognized while the prerogatives of science are preserved.

What is most interesting nowadays is that there is a resurgence of interest in this type of approach to the role of chance in the emergence of new phenomena. Indeed, after a long reign of mechanistic determinism or of its probabilistic variants, the recognition that chance has some objective reality is gaining ground again. Within physics the followers of Niels Bohr are gaining support against an Einsteinian view. But this movement of thought extends beyond the natural sciences and into the human ones (Dupuy 1982). In this transfer from the natural to the human sciences, thermodynamics provides many of the basic metaphors or methods. Authors such as Von Foerster (1960), Atlan (1979), Prigogine & Stengers (1979) are arguing that minor changes may produce considerable effects. Most noteworthy is the argument that a chance phenomenon can produce a fundamental rearrangement of a given system. The system, then, is able to reorganize itself and, hence one can speak of a self-organizing system. This legitimates a systems theory based upon thermodynamic interpretations because it lays the grounds for a non-mechanistic, non-deterministic, but systemic, approach to the relationships of a geographical entity to its environment (Phipps 1987). As was the case at the turn of the century, the conception of the role of chance becomes again a fundamental issue which makes research orientations differ among themselves.

The thermodynamic view establishes some role for chance, which may well express human initiative – whether it is individual or collective – as the promoter of new organizations. A (broadly understood) phenomenology is then in order to identify the ideology which justifies and underlies goal-oriented behavior. In so doing, one can reveal the new meaning structure which is promoting a new spatial arrangement, a new place for the unfolding of human agency (Berdoulay 1985a). In the same spirit, the study of utopias or myths as well as the evolution of collective identity may serve a similar purpose, (Morissonneau 1978, Claval 1980, Bélanger 1982, Bureau 1984).

The present-day concern for the local level, considered as place, resumes here all its significance because it can be approached within these theoretical perspectives. In the context of society as a whole, it expresses a quest for meaning. This is particularly clear, nowadays, when the welfare-state is in crisis and when there is a widespread dissatisfaction with consumption as the only yardstick of the individual's quality of life. After all, does not meaning come from an exercise in the reduction of environmental complexity, as shown by Turco (1982)? Facing the meaningless complexity of state and social organization (i.e. what Barel 1984 calls social emptiness), new territories, new places are wished for, and some are constructed. These are the new places which seem to be emerging at a meso-scale, such as in the contemporary European regionalism (Claval 1979). At a simple methodological level, one realizes the worthlessness of focusing exclusively upon either the individual actor or the socio-political structure. The actor is bound in a meaning structure which is local to a great extent. In the crisis of the welfare-state, the understanding of place and of its significance for the individual becomes imperative (Rosenvallon 1981).

One is led to think anew the approach to spatial planning (often tied to geography in French-language countries). In the planning process, the essential source of meaning has traditionally been considered to lie in the higher decision-making bodies, because they are supposed to be in tune with the modernization of society. We need not rehearse the disasters which this perspective has often produced at the local level. It often becomes impossible to continue talking of "place" after planners have done their work. Contrary to conventional planning wisdom the major source of meaning is, and must be, at the local level, the existing place. The problem is then to induce change such that the local meaning structure is respected. Obviously it has to evolve and, possibly, it may become totally new. In other words, place, as a system, has to deal with "noise" coming from higher organizational tiers. It may adapt either by integrating such "noise" within its own meaning structure (i.e., without changing it) or by reorganizing this structure in order to meet the disruptive tendencies which noise is bringing (Soubeyran & Barnier 1985).

In metaphorical terms at least, we see how place is a system of meaning which has to reorganize itself if some noise (whether planned or not) reveals itself to be too disruptive. In this case, noise (like chance) is producing the emergence of a new place. What is important here is that place operates like an autonomous entity, capable of changing its constitutive organization. From all this, one is thus led to conclude that it is necessary to study place with particular attention to its inner capacity to produce meaning, and not only with a search for its outside determinants. However, the question is left open as to how the meaning potential of place takes form and becomes actualized in the people's, or observers', minds. What follows is largely a personal perspective on the matter.

The necessity of discourse

We have seen from the above that to study place is to delineate its meaning, to be sensitive to its fundamental arrangement of elements and to be attentive to its autonomous capacity for producing meaning. In other words, a geographic account of place is like a whole staging process whereby people, objects, and messages are coordinated. It is like telling a story. My point is that the study of place has a strong narrative component. This may be said for geography in general (Berdoulay 1988), but it applies particularly well to the study of place. Obviously, this is not like telling any story. It has to reflect the actual interweaving of the relationships among those people, objects, and messages, which produces place and which may be viewed as a discourse.

How much, then, of this place narrative is a product of the geographer-observer or a product of its own reality? In order to cast some light on this fundamental issue, one must look more closely at discourse. The question becomes: is there a correspondence between the discourse on place by the analyst and the place-related discourse by the people, and vice versa? Let us look in turn at each one of them. Then, I will comment on the extent of their relationship.

The geographer's discourse has a number of modalities, which all have strengths or pitfalls. For example, metaphor is a key discursive procedure to express and promote radically new insights (Berdoulay 1982). It is much more than a comparison or an analogy, which it incorporates. It addresses itself to emotion by calling on one's imagination. Its aim is not explanatory; rather it is a means to suggest a new organization of our knowledge or information. It is an intuitive first step toward reconceptualization (which it is not), a powerful thought process leading to innovation. For many geographers, the organismic metaphor was just this, a means to introduce a new view of wholes, whereby one could envision the unity of the machine-like functioning of geographical entities and their ends. It was a way to integrate meaning and the then usual scientific procedures.

Geography, like all sciences, has a rhetorical dimension (cf. Schlanger 1975, Berdoulay 1988). Its aim is to formulate ideas, express them and convince. In order to succeed, it must reach emotion. And in this respect, the metaphor is particularly fit, because it relies on social or individual cognitive representations. Geographers' resort to this figure of speech opens their scientific discourse to the exterior world or to the unconscious. But, as a matter of fact, are not those mainsprings of meaning those which make explicit what a place *is* for the people?

Even in informal ways and about trivial matters, vernacular discourse about geographical space is widespread. It can be elicited and it is possible to show its own logic, as A. Gilbert (1985, 1986) has done about Quebec-city by using the reader's mail of a major newspaper. This discourse reveals how the environment is cognized and used in order to argue in favor of diverse issues. Affective bonds to place may become apparent. It is also possible to identify a whole rhetoric about place by its inhabitants (Lévy 1985). Metaphor, metonymy, synecdoche, etc., all serve the purpose of conceptualizing place and its meaning. These discourses express people's relation to their representations of the environment and society. The discourse on place is underlined by a discourse on social issues. They share a similar finality.

In other words, a place comes explicitly into being in the discourse of its inhabitants, and particularly in the rhetoric it promotes. Thus, the geographer's discourse uses the same ways as the people who define their own place. This is reinforced when there is a strong societal consensus to set this discourse in a highly coded landscape, as in Japan (Berque 1982). The geographer's discourse and the people's discourse are like two sides of the same coin. The similarity of their procedures points at the very nature of place. The study of this notion must entail its discursive, and particularly narrative, dimension.

In order to follow on this line of thought, let us turn again our attention to landscape as conceptualized above. This concept incorporates different environmental features which, taken one by one, are the result of some independent logic (be it economic, cultural, biological, etc.). This is why the landscape can be considered to be an autonomous level of spatial organization of these features, or, in other words, a transactional space where these different logics intersect. It is easy to recognize – from a semiotic point of view – that it is an "intertext", that is an organized ensemble of fragments of various texts which had been produced according to different logics (Kristeva 1978). What is important here is to note that the landscape is precisely where such new meaning structures are built. They are not a projection from outside logic; rather, they come about from all the spatial relationships among the various environmental features which originated in different realms (economic, political, biological, etc.). It is in this respect that movement, especially human movement, which connects various features, is most important. It particularly generates different perceptions of the visual environment (Berdoulay, 1985). Then, the place has a narrative level, where the meaning potential laying in the textual (landscape) level becomes actualized, or represented, for communicative purposes.

Such a view is useful on several counts. First, it ties together the insights provided on place by the concepts of lived space, and of landscape as transactional space. Second, it underlines the discursive nature of place which is at the source of its meaning. Third, it brings back the subject into place, as the former is largely being produced by the discursive process internal to place. Subject and landscape are constantly transforming each other (as noted by Soubeyran 1985). Fourth, it follows that the people as well as scientists share the same narrative task and tools for actualizing what place is about.

Thus place, as discourse, takes on a fundamental importance in geographic analysis and theory. As a narrative, it reminds us that our scientific understanding of it cannot be artificially separated from people's accounts of their place.

Conclusion

At the end of these personal reflections about place drawn from a selective reading of French language literature, let me recall the major points which may be of general interest. First of all, in order to make place a useful concept, one should consider that it is about meaning. Meaning, in turn, is grounded in social relations, but it unfolds within place, thanks to the autonomous level of landscape organization. Place is where meaning is largely being constructed, instead of simply being considered as projected from outside. This is why place is discourse. Subjectivity, then, is best viewed as part of the place-construction process. However, in the end, place is actualized at the narrative level – whether it is considered in a scientific way or not.

The view of place, which is argued for here, calls for important changes in our customary epistemological approach to landscapes and regions. Place as discourse opens the way for disturbing our persistent conceptual categories in order to fully consider meaning as a geographic process. After all is not *lieu* (or *topos*) also a term of Aristotelian rhetoric, a form in which a plurality of enunciations and reasonings coincide?

References

Andrews, H. F. 1984. L'oeuvre de Vidal de la Blache: notes bibliographiques. *Canadian Geographer* **28** (1), pp. 1–17.

Atlan, H. 1979. *Entre le cristal et la fumée*. Paris: Seuil.

Bailly, A. & J.-P. Ferrier 1986. Savoir lire le territoire: plaidoyer pour une géographie regionale attentive à la vie quotidienne. *Espace géographique* **15**, (4), pp. 259-64.

Barel, Y. 1984. *La société du vide*. Paris: Seuil.

Bélanger, M. 1982. "Conscientiser" la territorialité. *Bulletin de la Société neuchâteloise de géographie* **27**, pp. 389-404.

Berdoulay, V. 1981. *La formation de 1'école française de géographie (1870-1914)*. Paris: Bibliotheque Nationale (C.T.H.S., Memoires de la Section de Geographie, 11).

Berdoulay, V. 1982. La métaphore organiciste. Contribution à l'étude du langage des géographes. *Annales de géographie* 91 (507), pp. 573-86.

Berdoulay, V. 1983. Perspectivas actuales sobre el posibilismo: de Vidal de la Blache a la ciencia contemporánea. *Geo-Critica* **47**, pp. 5-27.

Berdoulay, V. 1985a. Les idéologies comme phénomènes géographiques. *Les Cahiers de Géographie du Québec* **29**, (77), pp. 205-16.

Berdoulay, V. 1985b. Convergences des analyses sémiotiques et écologiques du paysage. In *Paysage et Système. De l'organisation écologique l'organisation visuelle*, V. Berdoulay & M. Phipps (eds), pp. 141–53. Ottawa: University of Ottawa Press.

Berdoulay, V. 1988. *Des mots et des lieux. La dynamique du discours géographique.* Paris: Ed. du C.N.R.S.

Beroutchachvili, N. & J-L. Mathieu 1977. L'éthologie des géosystèmes. *Espace géographique* **6** (2), pp. 73-84.

Berque, A. 1982. *Vivre l'espace au Japon.* Paris: PUF.

Berr, H. 1903. Introduction général: la synthse des études relatives aux régions de France. *Revue de synthèse historique* **6**, pp. 166-81.

Bertrand, G. 1968. Paysage et géographie physique globale. Esquisse méthodologique. *Revue géographique des Pyrénées et du Sud-Ouest* **39**, (3), pp. 249-72.

Bloch, M. 1931. *Les caractères originaux de l'histoire rurale française.* Oslo: Institut pour l'etude comparée des civilisations.

Bonnemaison, J. 1986. *La dernière île.* Paris: Arlea-ORSTOM.

Braudel, F. 1949. *La Méditerranée et le monde méditerranéen à l'epoque de Philippe II.* Paris: A. Colin.

Braudel, F. 1986. *L'Identité de la France, I. Espace et histoire.* Paris: Arthaud-Flammarion.

Brossard, T. & J.-C. Wieber 1980. Essai de formulation systématique d'un mode d'approche du Paysage. *Bulletin de l'Association des Géographes Français* (468-9), pp. 103-111.

Brunet, R. 1969. Le quartier rural, structure régionale. *Revue Géographique des Pyrénées et du Sud-Ouest* (1), pp. 81-100.

Brunet, R. 1972. Pour une théorie de la géographie régionale. *La pensée géographique française contemporaine. Mélanges offerts à André Meynier.* Saint-Brieuc: Presses universitaires de Bretagne.

Brunet, R. 1980. La composition des modèles dans l'analyse spatiale. *Espace géographique* **9**, (4), pp. 253-65.

Brunhes, J. 1910. *La géographie humaine.* Paris: Alcan.

Bureau, L. 1984. *Entre l'Eden et l'utopie.* Montréal: Québec/Amérique.

Burgel, G. 1986. La ville, excès de société ou abus de langage. *Espaces temps* **33**, pp. 59-68.

Buttimer, A. 1971. *Society and milieu in the French geographic tradition.* Chicago: Rand McNally (AAG Monograph, 6).

Buttimer, A. 1978. Charism and context: the challenge of "la géographie humaine". In *Humanistic Geography*, D. Ley & M. Samuels (Eds), pp. 58-76. Chicago, Maaroufa Press.

Claval, P. 1973. *Principes de géographie sociale.* Paris: M.-Th. Genin.

Claval, P. 1978. *Espace et pouvoir.* Paris: PUF.

Claval, P. 1979. Régionalisme et consommation culturelle. *Espace géographique* **8**, (4), pp. 293-302.

Claval, P. 1980. Le québec et les idéologies territoriales. *Cahiers de Géographie du Québec* **24**, (61), pp. 31-46.

Claval, P. 1986. Les géographes français et le monde méditerranéen. IGU Congress. Barcelona.

Claval, P. & J.-P. Nardy 1968. *Pour le cinquantenaire de la mort de Paul Vidal de la Blache.* Paris: Les Belles Lettres.

Copeta, C. (ed.) 1986. *Esistere e abitare. Prospettive umanistiche nella geografia franco-fona.* Milano: Franco Angeli.

Coutau-Bégarie, H. 1983. *Le phènomène "Nouvelle histoire",* Paris: Economica.

Dardel, D. 1952. *L'Homme et la terre. Nature de la réalité géographique.* Paris: PUF.

Demangeon, A. 1905. *La plaine picarde et les régions voisines.* Paris: A. Colin.

Demangeon, A. 1923. *L'Empire britannique.* Paris: A. Colin.

Dion, R. 1934. *Essai sur la formation du paysage rural français.* Tours: Arrault.

Dupuy, J.-P. 1982. *Ordres et désordres.* Paris: Seuil.

Duby, G. & A. Wallon (eds.) 1975. *Histoire de la France rurale.* 2 vols. Paris: Seuil.

Febvre, L. 1905. La Franche-Comté. *Revue de synthèse historique* **10**, pp. 176–93, 319–42, and **11**, pp. 64–93.

Febvre, L. 1922. *La terre et l'évolution humaine.* Paris: A. Michel.

Ferrier, J.-P. 1984. *Antée 1.* Aix-en-Provence: Edisud.

Frémont, A. 1974. Recherches sur l'espace vécu. *Espace géographique* 3(3), pp. 231-238.

Frémont, A. 1979. Marginalité et espace vécu. In *Deux siècles de géographie francais* P. Pinchemel *et al.* (eds), pp. 336-43. Paris: CTHS.

Frémont, A., J. Gallais, J. Chevalier, M.-J. Bertramd & A. Metton 1982. *Espaces vécus et civilisations.* Paris: Ed. du G.N.R.S.

Frémont, A., J. Chevalier, R. Hérin & J. Renard 1984. *Géographie sociale.* Paris: Masson.

Gallais, J. 1976. De quelques aspects de l'espace vécu dans les civilisations du monde tropical. *L'Espace géographique* **5** (1), pp. 5-10.

Gallois, L. 1908. *Régions actuelles et noms de pays.* Paris: A. Colin.

Garcia Ballesteros, A. 1983. Vidal de la Blache en la crítica al neopositivismo en geografía. *Anales de Geografía de la Universidad Complutense* 3, pp. 25-31.

George, P. 1967. Le temps géographique. *Cahiers de Géographie du Québec* **11** (24), pp. 469-77.

Gilbert, A. 1985. Villes représentations collectives de l'espace et identité québécoise. *Cahiers de géographie du Québec* **29** (78), pp. 365-81.

Gilbert, A. 1986. L'analyse de contenu des discours sur l'espace: une méthode. *Géographe Canadien* **30** (1), pp. 13-25.

Grau, R. 1977. Sobre la base filosòfica del método regional en Vidal de la Blache. *V Coloquio de geografía,* Granada, pp. 297-301.

Gregory, D. 1981. Human agency and human geography. *Transactions, Institute of British Geographers* **6**, pp. 1-18.

Greimas, A.J. 1976. *Sémiotique et sciences sociales.* Paris: Seuil.

Kristeva, J. 1978. *Sèmeiòtikè: recherches pour une sémanalyse.* Paris: Seuil (1st ed 1969).

Lacoste, Y. 1982. *La géographie, ça sert, d'abord, à faire la guerre.* Paris: Maspero.

Le Roy Ladurie, E. 1966. *Les paysans de Languedoc.* 2 vols. Paris: SEVPEN.

Lévy, J. 1985. Des citadins contre la ville. Figures décalées, espaces refusés. *Hégoa* **1**, pp. 273-89.

Ley, D. 1977. Social geography and the taken-for-granted world. *Transactions of the Institute of British Geographers* n.s., **2**(4), pp. 498-512.

Ley, D. & M. Samuels, (eds) 1978. *Humanistic Geography.* Chicago: Maaroufa Press.

Lukermann, F. 1965. The "calcul des probabilités" and the "Ecole française de géographie". *Canadian Geographer* **9**, pp. 128-37.

Mann, H.-D. 1971. *Lucien Febvre. La pensée vivante d'un historien.* Paris: A. Colin.

Morissonneau, C. 1978. *La terre promise: le mythe du Nord québécois.* Montréal: Hurtubise HMH.

Nozawa, H. 1984. L'école géographique française et l'école des Annales. In *Languages, Paradigms and Schools in Geography*, K. Takeuchi (ed.), pp. 101-12. Tokyo, Hitotsubashi University (Lab. of Social Geography).

Phipps, M. 1985. Théorie de l'information et problématique du paysage. In *Paysage et système. De l'organisation écologique à l'organisation visuelle*, V. Berdoulay & M. Phipps (eds), pp. 59-74. Ottawa: University of Ottawa Press.

Phipps, M. 1987. Un mauvais adversaire pour un bon combat: un commentaire sur "l'analyse des systèmes en géographie humaine". *Canadian Geographer* **31**(1), pp. 47-9.

Phipps, M. & V. Berdoulay (eds) 1985. Paysage, système, organisation. In *Paysage et système. De l'organisation écologique l'organisation visuelle*, 9-19. Ottawa: University of Ottawa Press.

Pinchemel, P. 1970-72. Paul Vidal de la Blache. *Geographisches Taschenbuch*, pp. 266-79.

Prigogine, I. & I. Stengers 1979. *La nouvelle alliance*. Paris: Gallimard.

Raffestin, C. 1977. Paysage et territorialité. *Cahiers de Géographie du Québec* **21** (53-4), pp. 123-34.

Raffestin, C. 1980. *Pour une géographie du pouvoir*. Paris: Litec.

Raison, J.-P. 1976. Espaces significatifs et perspectives régionales à Madagascar. *L'Espace géographique* **5** (3), pp. 189-203.

Reynaud, A. 1981. *Société, espace et justice*. Paris: PUF.

Ritchot, G. 1985. Prémisses pour une théorie de la forme urbaine. In *Forme urbaine et pratique sociale*, G. Ritchot & C. Feltz (eds), pp. 23-65. Montréal: Le Préambule, and Louvain La Neuve, Editions Ciaco.

Rosenvallon, P. 1981. *La crise de l'Etat-Providence*. Paris: Seuil.

Roupnel, G. 1932. *Histoire de la campagne française*. Paris: Grasset.

Schlanger, J. 1975. *Penser la bouche pleine*. Paris: Mouton.

Soubeyran, O. 1985. Organisation, perception et émergence du sens dans le paysage. In *Paysage et système. De l'organisation écologique à l'organisation visualle*, V. Berdoulay & M. Phipps, (eds), pp. 155-66. Ottawa: University of Ottawa Press.

Soubeyran, O. & V. Barnier 1985. Les enjeux du virage aménagiste. *Loisir et société* **8**(1), pp. 55-91.

Turco, A. 1982. Le sens: est-il un concept pertinent en géographie de la perception? *Bulletin de la Société Neuchâteloise de Géographie* **27**, pp. 361-87.

Vant, A. 1981. *Imagerie et urbanisation. Recherches sur l'exemple stéphanois*. Saint-Etienne: Centre d'études foréziennes.

Vidal de la Blache, P. 1888-89. Des distributions fondamentales du sol français. *Bulletin littéraire* **2**, pp. 1-7, 49-57.

Vidal de la Blache, P. 1903. *Tableau géographique de la France*. Paris: Hachette.

Vidal de la Blache, P. 1917. *La France de l'Est (Lorraine-Alsace)*. Paris: A. Colin.

Villeneuve, P. 1981. La ville de Québec comme lieu de continuité, *Cahiers de géographie du Québec* **25** (64), pp. 49-60.

Von Foerster, H. 1960. On self-organizing systems and their environments. In *Self-organizing systems*, M.C. Yovitz & S. Cameron (eds), pp. 31-50. New York: Pergamon Press.

9

Place and culture: two disciplines, two concepts, two images of Christ, and a single goal

MILES RICHARDSON

The disciplines

As both Mikesell (1967) and Grossman (1977) note in their valuable reviews, geography and anthropology share several attributes. One attribute that separates them from such single-minded pursuits as geology or sociology is that both draw from diverse fields. The earth science interest of the physical geographer is matched by the life science interest of the physical anthropologist, the lure of the statistical survey of social science calls to both perception geographer and applied anthropologist, and the attraction of the landscape geographer to the beguiling world of the humanities finds a counterpart among humanistic anthropologists.

The diversity of interests presents to each discipline the challenge of maintaining unity, and each meets this challenge through celebrating the fieldwork experience. The explorer figure is part of the history, often mythic, of each. Both, for example, claim descent from such legendary men as John Wesley Powell, the self-taught naturalist, who lost an arm in the Battle of Shiloh and who, in addition to being the director of the Bureau of American Ethnology and of the United States Geological Survey, was the first European to navigate the Colorado River (Darrah 1951). Today, graduate students in geography and anthropology travel to the far corners of Asia, Africa, or Latin America, and accept as a fact of life Montezuma's Revenge – or the Old World variant, the Pharaoh's Curse – so they may collect plant specimens and phonemes for dissertations. Doctorate in hand, they join senior colleagues in such groups as the Conference of Latin Americanist Geographers or the Society of Latin American Anthropology where they modestly disclaim the harsh trials they have endured.

While common traditions of being in the field provide an experiential unity, an intellectual unity the two share is German scholarship. Although Alexander von Humboldt does not occupy the same exalted status in anthropology that he does in geography, his brother, Wilhelm, is lauded for his pioneering contributions to linguistics (Robins 1967, pp.174-8). Friedrich Ratzel, who coined the term *Anthropogeographie* to title his study of the manner in which different people interacted with different environments, is a common ancestor (Lowie 1937, pp. 119-27). Another German scholar, Franz Boas, who, according to his admirers, found American anthropology amateurish and anecdotal but left it a science (Benedict 1943, p.61), began his career studying the physical environment of the Arctic. He wrote in 1887 (Boas 1940), an essay, "The Study of Geography," in which he saw the field to be more a descriptive, historical science than a physical one, a position in line with his philosophical mentor, Wilhelm Dilthey (Harris 1968, pp.268-9).

Two descendants of the German intellectual heritage, Carl Sauer and Alfred Kroeber rose to the heights of their respective professions at the University of California. A student of theirs, and a man of many parts, Fred Kniffen brought the diversity of interest, the tradition of fieldwork, and the standards of scholarship to Louisiana State University, and there the two disciplines have intermingled to a degree unmatched elsewhere. Today, the Department of Geography and Anthropology continues to promote the exchange in pursuit of a common goal.

What is the common goal these two diversified, field- oriented, scholarly disciplines pursue?

Mikesell and Grossman both saw a commonality in the effort to specify the relationship between culture and environment. Clearly, any endeavor that includes within its makeup such a large component of the physical and natural sciences as do geography and anthropology will of necessity address nature. Equally clearly, any endeavor that includes such a mixture of social science and the humanities will address nature in a questioning mode. Estyn Evans, an historical geographer who, nonetheless, believes in the unity of his profession, posed that question in an address to the Department of Geography and Anthropology during a visit in 1969. The question that we ask, Evans suggested, is: What is man's place in nature?

The proposal of the geographer Evans, as it was earlier put by the anatomist, Thomas Henry Huxley (1897), and again by the philosopher, Max Scheler (1961), requires that we ask of stars in the heavens, of animals in the wilds, but most of all of ourselves in our solitude, where do we stand in nature's scheme? The question moves humanity to center stage and is, with all unabashed immodesty, anthropocentric. Consequently, it is a question that is asked not of an earth science, nor of a social science, nor a cognitive science, but of a human science.

The Concepts

The search for the answer to the question of where we are in nature's scheme begins with a consideration of two concepts that constitute the unique dialectic of being human and consequently concepts central to geography and anthropology, place and culture.

Place: We humans are flesh and blood primates and the world of the living and the dying is our home. As flesh and blood primates, we occupy space; as creatures of the symbol we transform that space into place.

The transformation of space into place parallels the other conversions we humans bring about through our work: stone into tool, behavior into conduct, sound into word, and, in a more sweeping way, nature into culture. The manner in which we achieve the conversion remains mysterious. Urged forward by the rise of field studies among the non-human primates there has been in anthropology since the early 1960s a rethinking of our primate nature. The rethinking has ranged from the sweeping generalizations about territorial imperatives to the more biologically informed concepts of sociobiology (Wilson 1975; Richardson 1978). While the rethinking, because of its reductionist nature, has yet to clarify the holistic magic of the conversion of nature into culture, it allows us to underscore how compelled we humans are to create, to symbolize, to speak. Likewise, it opens our eyes to the manner in which we blind ourselves to the inevitable reversion of tool to stone, conduct to behavior, word to sound, and, again in a more sweeping way, culture to nature.

Despite that we read in a recent issue of the *American Anthropologist* that "space becomes place through the implantation of people and events in the creation of a historically crafted landscape" (Rodman 1985, p. 68) and although we remember the important insights of Edward Hall (1969, 1977) who, in his insistence that space speaks, made the term "proxemics" an everyday word in the vocabulary of social science, place remains an undeveloped concept in anthropology. No doubt people "fascinate anthropologists more than location" (Robertson 1978, p.23), yet the reason why this important concept has not attracted the theoretical attention of anthropology goes back to the definition of culture as proposed by anthropology's 19th century progenitor, Edward B. Tylor: "Culture...is that complex whole which includes knowledge, belief, art, morals, law, custom, and any other capabilities and habits acquired by man as a member of society" (1958, p.1). The great contribution of Tylor was the establishment of culture as the intellectual property of people everywhere. Even the most primitive groups in the state of "savagery" had culture, and furthermore, their culture, like that of civilized society, was the product of systematized thought. Since people everywhere had culture, and since culture was a product of thinking, the specific places societies inhabited had little bearing on general cultural development. The dynamic of culture lay in what Tylor and his American counterpart Louis Morgan (1963) refer to as the laws of thought. The concept was place free.

The focus on culture as a thing in itself continued into the 20th century. Franz Boas, true to the German idealism of Dilthey, saw anthropology not as a universal, evolutionary science, but as a descriptive historical one, and under his guidance and numerous students, culture became cultures, and the key to the development of cultures lay not in the universal laws of thought but in the particular, immediate past. From this pluralization of culture arose the concept of culture area.

Developed by Boas' younger contemporary, Clark Wissler (1923) and promoted by Kroeber (1953), Boas' most famous student, the concept of culture area emerged from the study of the spatial distribution of cultures. Under the prevailing notions of scientific rigor, cultures were conceived as reducible to a series of observable units, or traits. This objectified enumeration permitted the tracing of individual traits, or larger trait-complexes, across space. The observed clustering of traits in a contiguous geographical area allowed the designation of that region as a culture area. In sum, although the concept of culture area produced regions of cultural uniqueness, the concept had little similarity with the more place-oriented *genre de vie* of Vidal de la Blache (1926). The concept was a purely cultural one and again, in essence, place free.

In contrast to anthropology, place in geography is a central concern. Indeed, place in the sense of "why things are where they are" is synonymous in the minds of many with the rise of geography as an academic discipline (James & Martin 1981), and place as a "lived-in" reality apparently goes back to the *pays* of Vidal de la Blache (Buttimer 1971, p.44). More recently, as an aftermath of the quantitative revolution which sought to make geography a spatial science and against any lingering desire to make it an earth science, place has evoked a rich exchange among those who would make geography an anthropocentric science.

As a human science, geography views the landscape not from an overhead, vertical angle in which the landscape is a value- free object, distant from the eye, but rather from the side, where the landscape is moral and the viewer located within (Tuan 1979). By integrating felt experience with the discerning symbol, the humanistic geographer prepares himself to offer an interpretation of place in its ambiguity, ambivalence, and complexity (Tuan 1973, 1976).

Drawing upon phenomenology, an anthropocentric geography attacks both the conventional science distinction between the perceiving subject and the perceived object and the idealism that transforms nature into a cognitive map (Buttimer 1976). Consequently, place is not simply a passive mirror that dispassionately reflects the human endeavor, but it is the medium, the gesturing (Hugill 1984), through which we are (Rowntree & Conkey 1980).

In this exchange, humanistic geographers have established the centrality of we, *homo creator* (Laan & Piersma 1982), the intersubjectivity of the created landscape, and the reciprocity of creator and artifact to shape an emergent ontology. In the most ambitious treatment to date, Pred (1984) has sought to integrate creator, artifact, and emergent becoming into a temporal process that includes social structure and physical nature. Place is a "historically contingent process" in which biography and structure reshape themselves into one another accompanied by a similar cease-less metamorphic cycling between nature and artifact; consequently, the "transformation of the physical world is inseparable from the becoming of place" (Pred 1984, p. 287).

Culture

To turn the world of nature, a world in which we breathe, into the world of symbols, a world in which we speak, is the special destiny of us humans. Culture, the second of the two core concepts, is the transfiguration of that destiny.

In geography, culture is both the subject and the explanatory tool of cultural geography. As developed by Kniffen and other members of the "Berkeley" school, cultural geography concentrates on the landscapes of small scale, folk, or rural societies. Indeed, some in cultural geography (Denevan 1983) would restrict by preference their field to the artifactual imprint the members of these groups have left on the landscape: the Anglo-American dogtrot house, the Upland South scraped cemetery, or the flat-roofed dwelling of rural Mexico. In addition to being a subject, culture is offered as the explanation of a landscape. A landscape has the characteristics that it has because of the culture that inhabits that territory, and that particular culture has its particularities because of its history. The concept comes directly from the historical school of Boas, Kroeber, and Wissler. By using the concept of culture as a series of more or less discrete traits, the cultural geographer could approach the subject of distinctive landscapes systematically, and in contrast to the holistic, and therefore to a degree, "intuitive" methods of the regionalist, objectively trace the distribution of traits so as to discover their origins.

In the hands of the master, cultural geography remained tied to the earth. Behind the artifacts, people, however dimly glimpsed, spread across the land, tending their livestock in distinctive barns, bounding their fields, green with crops borrowed from indigenous cultures, with distinctive fences, and burying their dead in ground made hallow by grief. Notwithstanding the master's touch, the diffusionist's concept of culture as discrete, objective traits constantly threaten to achieve a reality *sui generus* and to become, through the magic of reification, a superorganism. On that occasion, the concept that once illuminated particularity, becomes one that hides it, and thus warrants criticism (Duncan 1980).

Culture is to anthropology what place is to geography, a subject of considerable discussion. Tylor's famous definition, which recounts those qualities that separate us from other animals, continued to be cited well into the 20th century. With the rise of the Boasian school, however, the grand evolutionary scheme of the 19th century that chronicled the march of mankind from savagery to civilization was reduced to a more restricted (that is to say, "scientific") investigation into the history, or the diffusion, of traits. In Germany, where the logic of diffusionism routed evolutionary thought, the movement of traits and trait complex was conceived as occurring of world-wide basis, and in this, the *Kulturkrieslehre* of Fritz Graebner and Wilhelm Schmidt (Kluckholn 1936) may have foreshadowed the fascination some students of Sauer have for pre-Columbian, trans-Pacific contact. In the United States, where Boas was as critical of global diffusion (1911) as he was of global evolution, the emphasis on the diffusion of discrete traits had developed by 1930 into a concern for what those traits mean. Culture was moving from being a list to becoming an integrated whole.

The magnitude of the contrast between culture as a list and culture as an integrated whole is measured in the distinction between Wissler's definition of Plains Indian culture with that of Ruth Benedict's.

Wissler wrote that of 31 tribal groups inhabiting the great plains,

"eleven may be considering as manifesting the typical culture of the area. . . . The chief traits of this culture are the dependence upon the buffalo or bison and the very limited use of roots and berries; absence of fishing; lack of agriculture; the tipi as a movable dwelling; transportation by land only, with the dog and the travois (in historic times with the horse); want of basketry and pottery; no true weaving; clothing of buffalo and deerskins; a special bead technique; high development of work in skins; special rawhide work (parfleche, cylindrical bag, etc.); use of the circular shield; weak development of work in wood, stone, and bone. Their art is strongly geometric, but as a whole, not symbolic; social organization tends to simple band; a camp circle organization; a series of societies for men; sun dance ceremony; sweat house observances, scalp dances, etc." (1957, p.220-2).

Benedict came to anthropology from literature, and as a student of Boas in his seniority she articulated the historical particularist's concern for meaning. In her epochal *Patterns of Culture*, she wrote, "Cultures . . . are more than the sum of their traits. We may know all about the distribution of a tribe's form of marriage, ritual dances, and puberty initiations, and yet understand nothing of the culture as a whole which has used these elements to its own purpose" (1934, p.47). In achieving its own purpose a culture selects from the great arc of possibilities offered by the environment, by the human life cycle, and by the neighboring cultures those traits that fit its purpose and integrates them into a pattern. Among the Indians of the great plains, this pattern was permeated by a Dionysian desire to achieve the extraordinary. Hence mourning behavior was an

"indulgence in uninhibited grief. . . . The women gashed their heads and their legs, and cut off their fingers. Long lines...marched through the camp after the death of an important person, their legs bare and bleeding. . . . The possessions of the dead were not thought to be polluted, but all the property of the household was given away because in its grief the family could have no interest in things they owned and no use for them. Even the lodge was pulled down and given to another. Nothing was left to the widow but the blanket around her." (1934, p.112).

The contrast between Wissler's definition of a culture as a list of discrete, objective traits and Benedict's definition of it as a pattern into which traits are integrated to form a meaningful whole came to be the major distinction between cultural geography and cultural anthropology. While cultural geography continued to explain the distinctive landscapes of small scale societies as the product of diffusion of traits, cultural anthropology turned almost completely to the pursuit of patterns. Benedict's articulation of the concern for meaning, which she cites as derived from Dilthey's *Die Typen de Weltanschauung* (Benedict 1934, p.52), was so successful that it became conventional wisdom in anthropology. Since a culture forms an integrated whole, then for the anthropologist to understand any one trait, the anthropologist must see that trait in its relationship to the others. To accomplish this understanding the anthropologist must adopt the native's point of view. This Benedictine creed remains an article of faith in much of contemporary anthropology.

Benedict denied that to speak of the purpose of a culture was in any sense mystical, and she argued only the language-forms we are forced to write in make culture seem to have a consciousness (1934, p. 47-8). Despite her assertions, even in her own writing the concept of culture as a meaningful pattern carried with it the concept that the pattern, the integrated whole, exists at a level distinct from the individual. Culture became an it, and particular cultures, particular its; culture changed from being the creative genius of the species that allowed us to march from savagery to civilization to become a coercive hand that shapes us to fit its purpose. George Stocking, the authority on Boas, reviewed this Boasian-inspired mutation and concluded, "Almost unnoticed, the idea of culture, which once connoted all that freed man from the blind weight of tradition, was now identified with that very burden, and that burden was seen as functional to the continuing daily existence of individuals in any culture" (1966, p.878).

Coupled with the concept of social system, derived ultimately from the French sociologist, Emile Durkheim, the view of culture as a meaningful whole led quickly to a search to discover the linkage between culture and personality. Paralleling a similar interest in geography that a landscape, or a region, or even a country may have a personality, culture and personality flourished in the 1930s and 1940s. It continued after World War II in attenuated form, but beset with conceptual difficulties, not the least of which was the comparison between culture, an it, with personality, another it, culture and personality has all but disappeared into a more generalized psychological anthropology.

The conceptualization that culture exists at a level distinct from that of the individual and that the individual falls victim to its dispassionate sweep culminated separately in the writings of Kroeber and Leslie White. Both argued that sciences form a hierarchy with each science resting upon a more basic science but not reducible to it. Biology, for example, is dependent upon chemistry, but (as any biologist will insist) is not reducible to chemistry. Attempts to do so commit the fallacy of reductionism; the reduction process destroys the subject matter of the higher science. The aim of a "historical" science such as anthropology, to paraphrase Kroeber (1952, p.71), is to keep "the phenomena whole."

Kroeber, the senior scholar, first addressed the issue of keeping the phenomena whole in "The Superorganic," published in 1917; he returned to the matter throughout his long and productive career. His concept of culture as supra-psychal and supra-societal led him to describe culture as having patterns of development similar to style – like Benedict, he came into anthropology from literature. Cultures follow courses of experimentation, flowering, fatigue, and collapse. These are internal qualities of culture as such and are not reducible to economic causation. White, the more eclectic scholar, drew explicitly upon Durkheim and implicitly upon Karl Marx to argue culture is a system – not a pattern – and composed of the subsystems of technology, social relations, and ideology. A change in one subsystem, particularly the technological one, leads to changes in the others. Kroeber, who desired to be at once both integrative and particularizing, sought the idiographic genius of a culture; White, the generalizing scientist, sought nomothetic regularities.

Both Kroeber and White argued forthrightly for the priority of culture in determining change. "A hundred Aristotles among our cave-dwelling ancestors . . . would have contributed far less...than a dozen plodding mediocrities in the 20th century" (Kroeber 1952, p.42). "Warfare is a struggle between social organisms, not individuals. . . . To picture the multitudes of docile serfs and peasants of ancient Egypt, pre-Columbian Peru, China or Czarist Russia going to war because of an 'innate pugnacity and a lover of glory' . . . or . . . because 'men like war' is grotesque. They were forced to go, driven to slaughter like sheep" (White 1949, pp. 132-3). Both were charged with reifying, and in White's case, even deifying culture (for Kroeber, Sapir 1917; for White, Strong 1953).

In the defense of his position, White, in particular, presented a sophisticated argument. Anticipating a prevailing consensus among contemporary cultural anthropologists, White argued that the central aspect of culture is not that it is learned behavior, but that it is symbolic behavior, that is, things and events brought about by the human capacity to assign meaning. Things and events endowed with meaning, or "symbolates," may then be considered extra-somatically, as a superorganic system. Pointing to the commonly accepted division between *la langue* (language) and *la parole* (speech), White observed that linguists – even and especially those who accused him of reification – felt perfectly justified in analyzing language without recourse to the speaker. The intricacies of English grammar may be pondered without an appeal to the speaker's eye color, nose shape, digestive health, or belief in God. Similarly the anthropologist could, and as a scientist, he must, analyze culture as if it had an existence distinct from humans (White 1959, pp. 227-51).

The contributions of Julian Steward (1955), White's contemporary, are well-known in geography. Like Kniffen, Steward was a student of Kroeber's, but unlike Kniffen, Steward stressed not diffusion but adaptation as the cause of culture change. The distinction between cultural geography and cultural ecology led Grossman to characterize the emphasis in cultural geography as adaptation *of* the environment, in which man is viewed as a major agent on the landscape, and that in cultural ecology as adaptation *to* the environment, in which the environment is seen as having a creative role (Grossman 1977, p.132). The division, however, is becoming increasingly blurred (Denevan 1983).

The writings of Steward and White broke the grip of the Boasian paradigm of historical particularlism and ushered in the nomothetic revival, which flowered after World War II and which is still vigorously avowed by Marvin Harris (1980). In the 1950s, however, a rival concept of culture began to emerge.

The concept arose out of the traditional study in anthropology of the systematic, yet divergent manner in which different peoples classify those to whom they are kin. Employing techniques successfully utilized in linguistics to secure an observer-free description of the informant's language, proponents of the concept promised to introduce similar rigor to the ethnographic process, and consequently their proposal became known as the "new ethnography" – which followed the "new physical anthropology" but was antecedent to the "new archaeology." The concept lay claim also to the now traditional position in anthropology of seeing the world from the native's point of view. The goal was to elicit observer-free classifications and to offer rules that would account for the contrastive distribution and hierarchial ranking of component terms within semantic domains. Based on the similarity of the procedure to one that a linguist employs in securing a phonemic description, the procedure itself was considered to be an emic analysis. (Emic derives from phon*emic*, and etic, which is the imposition of the scientist's categories upon the data, from phon*etic*. Pike 1966.)

In conjunction with structuralism, the movement flowered under various guises, componential analysis, ethnoscience, ethnographic semantics, as it sought to discover the rules not only of kinship but also disease categories, color classifications, and environmental features. Eventually, the movement coalesced into cognitive anthropology.

As articulated by one of its foremost exponents, culture, in cognitive anthropology, "consists of whatever it is one has to know or believe in order to operate in a manner acceptable to its members. Culture is not a material phenomena; it does not consist of things, people, behavior, or emotions. It is the form of things that people have in mind. . . ." (Goodenough 1957, p.167). So conceived, culture achieves the epistemological status of language (Keesing 1974, p.77).

The claim that culture is mental forms and that these forms may be accessed through the rigorous application of observer-free procedures – "extreme subjectivism married with extreme formalism" (Geertz 1973, p.11) – parallels similar claims in the field of perception geography. In that field mental maps that individuals allegedly possess are accessed through objective interviews and questionnaires, the response to which are treated with sophisticated statistical techniques (Bunting & Guelke 1979). Both cognitive anthropology and perception geography argue their strength, apart from the scientific precision both say they control, is in the presentation of the world as classified or mapped by the native. Yet nowhere in the elegant cognitive structures of either do flesh and blood creatures appear. In these denatured, bleached accounts of the human endeavor, people are not martyrs to superorganic patterns or systems; they are not even there.

The pattern of the historical particularist and the system of the generalizing scientist turned culture into an it and located the it outside the individual organism. Cognitive anthropology turned culture into a structure and located that structure inside the individual's head. Despite the apparent gap between external pattern-system on the one hand and internal structure on the other, both arguments use a language concept of culture and both, in their own distinctive manner, juxtapose culture and the individual; in one the individual is a sacrificial victim, and in the other, a bloodless receptacle.

In the anthropology of the 1980s, as we move through the post-structural world and toward the post-modern era of the 21st century, we read more and more critiques such as these: "rather than providing a set of rules people must obey. . . , kinship provides an idiom by which people seek to maintain or transform their relationships to others as the situation demands" (Bledsoe 1980, p.29); the words we use in everyday discourse are "quasi- literary creations" and as such they "do not refer or reflect, they signify and constitute, that is, they objectify, [and] make real through symbolic means . . ." (Varenne 1978, pp. 646-7); and ". . . anthropologists may have greatly exaggerated the systemic aspects of beliefs. . ." for culture, "after all, is a continuous, creative, [and] inventive process . . ." (Crick 1982, pp. 295, 299). "A constituted cultural order continually comes into being and never exists as a *fait accompli.*" Furthermore, the ambiguity of the constitutive "is not simply something to be denied or explained away. . . . It is a fact of human experience that both inhibits and motivates much behavior" and must be evoked. Anthropology "is an evocative science not only in the sense of giving voice to our informants but in trying to evoke by means of their voices, and by other devices [such as place], the 'thickness' of human experience in culture" (Dougherty & Fernandez 1982, p.825, 827).

Culture, in these critiques, has moved "down" from its reified position as an external superorganism; it has moved "out" from its being an internal structure in the individual's head; and it now resides in "words for the most part, but also gestures, drawings, musical sounds, mechanical devices like clocks or natural objects like jewels. . . ," and "its natural habitat is the house yard, the market place, and the town square" (Geertz 1973, p. 45). It has become public, truly social, and intersubjective, for although "ideational, it does not exist in someone's head; though unphysical, it is not an occult entity" (Geertz 1973, p. 10). It has become, once again, what we, we humans, achieve.

This concept of culture aims to accomplish the Benedictine charge to see the world from the native's point of view. The native is neither victim nor receptacle, nor is the native a distant other with a bone through his nose, but you and I, man, woman, together, united in the human struggle to be. The native, however, is speaking his own story, the story of his own people, and again in affirmation of the Benedictine doctrine, we must preserve the sanctity of his text, even at the expense of our compulsion to edit his to fit yours or mine. At the same time, the story the native tells speaks to us in its humanity, and we must listen, even as he recites in gleeful tone the obscenities committed in the name of love.

This concept of culture, associated with interpretive or symbolic anthropology – the two principal spokesmen of which are Clifford Geertz (1973, 1983) and Victor Turner (1982, 1985) – is humanistic, and as such parallels the concept of place in anthropocentric geography. Pattern, system, and structure have been replaced with text, in the case of Geertz, and performance, in the case of Turner, and in both Geertz and Turner and in anthropology in general, the timeless, placeless language model of culture of earlier writers is giving way to a time-bound, place-specific, experience-rich, speech model of culture (Ricoeur 1979). Instead of opting for *la langue* as did White and the structural linguists, we opt for *la parole*.

In the speech, discourse model of culture, the speaker and the audience reciprocally address one another about the subject at hand. Language is subject free, but speech is event rich and always occurs in the context of situations. We address one another across the genetically fixed void that lies between us as you and me and not as generic *Homo sapiens* (Geertz 1973, p.49). We exist, you and I, in the dialogue of specific words, which have their meaning because of the specific culture they evoke, which in turn, comes to be in a particular place. (Richardson 1982; Richardson & Dunton 1988).

The images

Place, the experiential; culture, the symbolic. Place, the world; culture, the world view. Place, the "historically contingent process;" culture, the "creative," the "constitutive." Let us put the two core concepts of the 1980s together; let us join the two humanistic fields in the joint search for our place in nature; let us do so in two distinct places where two cultures come to be, the Catholic *iglesia* of Spanish America and the Baptist church of the American South.[1]

Not all *iglesias* in Spanish America, that vast region which begins somewhere *north* of the Rio Grande and extends to Tierra del Fuego, are massive, colonial cathedrals. Yet if we were to pick a typical *iglesia*, we would opt for one fronting the plaza, its huge doors opening directly on to the street. Along one side, a vendor sells pictures of the saints, booklets for novenas, and candles for offerings. On the other side, crouched close to the door so none may miss his festering sores, a beggar holds out his hand. Above the giant arms of the tower clock move forward, and higher still, the bells peel the beginning of mass.

Inside, the noise of the street, a bus revving up, a truck passing by, follows us as we kneel and cross ourselves. Along both sides saints peer down from their niches, their eyes alight from the candles' glow. At one, a woman and her daughter kneel; at another, a man reaches out to touch the statue's feet. Among the saints, the Virgin, her eyes filled with distant sadness, offers solace. Near the entrance, or in a chapel to one side, but commonly high above the altar, his thorn-crowned head slumped to one side, his pierced side bleeding, Christ hangs in perpetual agony.

Not all churches among the Southern Baptists, that almost exclusively white denomination and, with over 14 million members, the largest Protestant group in the United States, are new, modern structures. Yet if we were to pick a typical church, we would select one whose sign out front announces this is the First Baptist and whose location is not far from the county court-house. As we cross the parking lot to the church we read the name of the minister, Rev. Tommy Overton, and the title of his sermon, "Evolution? The Bible Says No!"

No vendors, no bells, and certainly no beggars greet us as we step into the air-conditioned comfort. No street noises compete for our attention as we sit on the cushioned pew. No statues and no candles adorn the plain walls. The carpeted floor slopes down front to the table used for the Lord's supper. On top is an opened Bible and nearby, the collection plates. The platform above the table supports not an altar but a massive oak pulpit strong enough to withstand the slam of Rev. Overton's fist. Behind the pulpit are seats even now being filled by the robed choir, the high-pitched ladies in front, the deeper-voiced men in the rear. Above the choir at the spot where in Spanish America, Christ hangs in perpetual agony, is the curtained baptistry. The curtain is closed now, but on the occasion when the congregation welcomes the newly saved into the bosom of its fellowship, the curtain is parted, and before the audience, Rev. Overton gently immerses the converted sinner so that he, like Christ, may rise from the tomb and live forever.

Two places, two cultures, and two Christs. If the common goal of geography and anthropology is to understand our place in nature, what is the immediate task? The immediate task before us is obviously not one of prediction; as humanistic scientists, ours is "not an experimental science in search of law but an interpretive one in search of meaning" (Geertz 1973, p.5). The search for meaning calls us to explicate the process whereby we construct meaning, whereby we speak reality into existence, speak with both hand and mouth. Benedictines still, we profess the task of explication from the native's point of view. Such an explication makes no claim to knowing what goes on in the native's head. The native's head belongs to the native; *A skeleton key to Finnegan's Wake*, Geertz reminds us (1973, p.15), is not *Finnegan's Wake*. Similarly, as anthropocentric scientists, we bracket the native's claims that his reality extends beyond the range of human achievement; yet our bracketing should be sympathetic and our terms, "user friendly."

The speech model of culture is a model of culture as speaking; it gains its strength from its view of culture as an ongoing creative discourse. Natives, on the other hand, especially Christian natives, report, and insist they have, a transcendence that exceeds the moment. As long as we are humanistic scientists, we must suspend their belief, and at the same time we must take that very belief seriously. To accomplish this paradox, we can do no better than to turn to the apostle Paul who preached that the Christian place was to be in Christ. "For as by a man came death, by a man has come also the resurrection of the dead. For as in Adam all die, so also in Christ shall all be made alive" I Corinthians 15: 21-22 (RSV). For Christians to be, they must be in Christ; to be Christians, they must have Christ to be in. Being-in-Christ requires, however, for a flesh and blood, symbol-cursed creature, a specific Christ to be in. In the same manner that we are both you and I before we are *Homo sapiens*, so Christians are, in their humanity, first, and perhaps foremost, Spanish American Catholics kneeling before a suffering Christ or Southern Baptists singing the glories of one who is risen.

To bring forth a Christ to be in, to create a specific consciousness of Christ, both Spanish American Catholics and Southern Baptists seek to escape the continually creative dynamic of the speech model of culture through reification (Richardson 1981). Reification, to make process into a thing, although a social science sin, is a common virtue of everyday life. From a constructionist's perspective (summarized in Thomason 1982), reification is the final step on a process, dictated by our genes, first to speak (more broadly, to symbolize), next to achieve consciousness of our speaking (to take the role of the other), and finally, and ultimately, to turn the word into flesh. In everyday life, reification makes social life possible; it allows us to gain knowledge that we exist, and existing as individual flesh and blood creatures, that we die. In religious life, reification makes transcendence possible, to exist beyond the moment and to deny death its victory.

In Spanish America, Catholic Christians deny death its victory through the creation of a suffering Christ, a sensuous Christ, a Christ of the touch. Christ, in his suffering, teaches us how to suffer, and through suffering in the knowledge of Christ, we transcend death. In the American South, Southern Baptists deny death its sting through the creation of a risen Christ, a spoken Christ, a Christ of the word. Christ, having risen from the tomb, teaches if we but speak his name in faith, we too will rise to be with him in heaven.

Catholics and Baptists create their specific Christs to be within two distinct places. In Spanish America, the place is the *iglesia*-temple, a place in itself made holy through ritual. In the sacred temple, candles, smoke, statues remind us that God sent his son to live among us as one of us, to suffer and to die. In the American South, the place is the church-auditorium, in itself a profane structure, where even the baptistry, the empty tomb, is filled with tap water. In the secular auditorium, the absence of noise and smell and of visual scenes reminds us that God has called the preacher to pulpit so that he may deliver the Holy Word of salvation through faith. In both places, the creative process is to deny death, to put death in its place.

Conclusion

In what must surely be one of the classics in social science literature, Ernest Becker (1973) wrote ". . . our central calling, our main task on this planet, is the heroic" (Becker 1973, p.1). Culture "everywhere is a living myth of the significance of human life, a defiant creation of meaning" (Becker 1973, p.7). The greatest threat to the creation of meaning is death. Consequently, we, you and I, no less than the Catholic in Spanish America or the Baptist in the American South, must conquer death. "Man is *Homo sapiens* precisely because he is a creature that needs to be 'in place'. . . ." (Houston 1978, p.227). Victory over death is to put death in its place. In everyday life victory over death is accomplished through assigning death to the *das Manh*, the they-self (Heidegger 1962, p.163-168). When we read that Jackie Gleason has died only a few days after Fred Astaire, we in the wisdom of the evening Martini remark to one another, "They say that everyone has to go sometime." In Christian life, victory over death is accomplished through the creation of a Christ figure, who himself dies and rises so that we may transcend our being.

Geography and anthropology, as humanistic sciences, committed both to the experiential as well as to the symbolic, must bracket both everyday and Christian life. Within these brackets we see not the myth of victory over death but the myth of Sisyphus. Sisyphus, who stole secrets from the gods, is condemned by them to push a boulder up a mountain, where, at the top, the boulder rolls back down to await the next struggle up to the top, where it rolls back again. "You have already grasped that Sisyphus is the absurd hero. He is, as much through his passions as through his torture. His scorn of the gods, his hatred of death, and his passion for life won him that unspeakable penalty in which the whole being is exerted toward accomplishing nothing" (Camus 1979, p.54). Absurd though Sisyphus may be, heroic he surely is. "The struggle itself toward the heights is enough to fill a man's heart" (Camus 1979, p.56).

Man's place in nature is to put death in its place, which is an effort that places man outside of nature, which is, of course, an impossibility, which makes it, therefore, a distinctly human task.

We, geographers and anthropologists, joining with John Wesley Powell, the one-armed hero, to navigate the Colorado on its tortuous course from mountain to gulf, search for understanding how it is a bipedal primate yearns to live forever. Our pursuit of this understanding will surely increase the burden of our awareness of where we humans stand in nature's scheme, a pursuit that promises, under the load of its burden, not efficiency, but a more precious commodity, wisdom (Tuan 1977, 1984).

Meine Bürde

Why are they there?
I asked the priest
about the rocks at
the foot of the cross
on the mountain.

The Indians carry them,
he replied. But why?
It has to do with their *pena*.

Their *pena*, their *pena*.
La pena, la pena suya,
la pena de los indios.
La pena of the past.

The load on a back,
head thrusted forward,
up the mountain,
and back down. *La pena.*
La pena suya; la pena nuestra.
Le mythe de Sisyphe.

153

Acknowledgments

Most of my academic life has been spent in the presence of geographers. I have taken classes from them, I have listened to them in the hallways and over coffee, and I have heard them argue with one another in meetings and read their debates in the Annals. I am richer from my associations. I especially appreciate the conversations with James S. Duncan, Nancy G. Duncan, and Jonathan Smith.

Note

1 The following descriptive paragraphs draw heavily upon Richardson 1986; the discussion on bracketing, reification, and Sisyphus, however, is unique to this essay.

References

Becker, E. 1973. *The denial of death*. New York: The Free Press.

Benedict, R. 1934. *Patterns of culture*. Boston: Houghton Mifflin.

Benedict, R. 1943. Obituary of Franz Boas. *Science* **97**: pp. 60-2.

Bledsoe, C. 1980. The manipulation of Kpelle social fatherhood. *Ethnology* **XIX**, pp. 1-28.

Boas, F. 1911. Review of Graebner: Methode der Ethnologie. *Science* **34**, pp. 804-10; Reprinted in Boas 1940, pp. 295-304.

Boas, F. 1940. *Race, language, and culture*. New York: Macmillan.

Bunting, T. E. & L. Guelke 1979. Behavioral and perception geography: a critical appraisal. *Annals of the Association of American Geographers* **69**, pp. 448-62.

Buttimer, A. 1971. *Society and milieu in the French geographic tradition*. Chicago: Rand McNally.

Buttimer, A. 1976. Grasping the dynamism of lifeworld. *Annals of the Association of American Geographers* **66**, pp. 277-92.

Crick, M. R. 1982. Anthropology of knowledge. *Annual Review of Anthropology* **11**, pp. 287-313.

Camus, A. 1979. *The essential writings*. Edited with interpretive essays by Robert E. Meagher. New York: Harper Colophon Books.

Darrah, W. C. 1951. *Powell of the Colorado*. Princeton: Princeton University Press.

Denevan, W. M. 1983. Adaptation, variation, and cultural geography. *The Professional Geographer* **35**, pp. 399-406.

Dougherty, J. W. D. & J. W. Fernandez 1982. Afterword. *American Ethnologist* **9**, pp. 820-32.

Duncan, J. S. 1980. The superorganic in American cultural geography. *Annals of the Association of American Geographers* **70**, pp. 181-98.

Geertz, C. 1973. *The interpretation of cultures*. New York: Basic Books.

Geertz, C. 1983. *Local knowledge*. New York: Basic Books.

Goodenough, W. 1957. Cultural anthropology and linguistics. In *Report of the Seventh Annual Round Table Meeting on Linguistics and Language Study*, P. Garvin (ed.). pp. 167-77. Washington DC: Georgetown University.

Grossman, L. 1977. Man-environment relationships in anthropology and geography. *Annals of the Association of American Geographers* **67**, pp. 126-44.

Hall, E. 1969. *The hidden dimension*. Garden City, NY: Anchor Books.

Hall, E. 1977. *Beyond culture*. Garden City, NY: Anchor Books.

Harris, M. 1968. *The rise of anthropological theory*. New York: Thomas Y. Crowell.

Harris, M. 1980. *Cultural materialism*. New York: Vintage Books.

Heidegger, M. 1962. *Being and time*. New York: Harper & Row.

Houston, J. M. 1978. The concepts of "place" and "land" in the Judaeo-Christian tradition. In *Humanistic geography: prospects and problems*, D. Ley & M. S. Samuels (eds). pp. 224-37. Chicago: Maaroufa Press.

Hugill, P. 1984. The landscape as a code for conduct. In *Place: experience and symbol*, M. Richardson (ed.). pp. 21-30 *Geoscience and Man* **24**.

Huxley, T. H. 1897. *Evidence of Man's place in nature*. New York: D. Appleton.

James, P. E. & G. Martin. 1981. *All possible worlds*. New York: Wiley.

Keesing, R. M. 1974. Theories of culture. *Annual Review of Anthropology* **3**, 73-97.

Kluckholn, C. 1936. Some reflections on the method and theory of the Kultur-kreislehre. *American Anthropologist* **38**, pp. 157-96.

Kroeber, A. L. 1952. *The nature of culture*. Chicago: University of Chicago Press.

Kroeber, A. L. 1953. *Cultural and natural areas of native North America*. Berkeley: University of California Press.

Laan, L. van de & A. Piersma 1982. The image of man: paradigmatic cornerstone in human geography. *Annals of the Association of American Geographers* **72**, pp. 411-26.

Lowie, R. H. 1937. *The history of ethnological theory*. New York: Holt, Rineheart and Winston.

Mikesell, M. 1967. Geographic perspectives in anthropology. *Annals of the Association of American Geographers* **57**, pp. 617-34.

Morgan, L. H. 1963. *Ancient society*. Cleveland, OH: World Publishing Co.

Pike, K. L. 1966. Etic and emic standpoints for the description of behavior. In *Communication and culture*, A. G. Smith (ed.), pp. 152-67. New York: Holt, Rinehart & Winston.

Pred, A. 1984. Place as historically contingent process: structuration and the time-geography of becoming places. *Annals of the Association of American Geographers* **74**, pp. 279-97.

Richardson, M. 1978. DNA or God? *The Anthropology and Humanism Quarterly* **3**, pp. 7-8.

Richardson, M. 1981. On "the superorganic in American cultural geography." *Annals of the Association of American Geographers* **71**, pp. 284-7.

Richardson, M. 1982. Being-in-the-market versus Being-in-the-plaza. *American Ethnologist* **9**, pp. 421-36.

Richardson, M. 1986. Material culture and being-in-Christ in Spanish America and the American south. In *Architecture and cultural change*, D. G. Saile (ed.) pp. 25-30. Lawrence, Kansas: School of Architecture and Urban Design.

Richardson, M. & R. Dunton 1988. Culture in its places: a humanistic perspective. In *The relevance of culture*, M. Freilich (ed.). In press.

Ricoeur, P. 1979. The model of the text: meaningful action considered as a text. In *Interpretive social science*, P. Rabinow & W. M. Sullivan (eds) pp. 73-102. Berkeley: University of California Press.

Robertson, D. 1978. A behavioral portrait of the Mexican plaza principal. PhD Dissertation, Department of Geography, Syracuse University. Syracuse, New York.

Robins, R. H. 1967. *A short history of linguistics*. Bloomington: Indiana University Press.

Rodman, M. C. 1985. Moving houses: residential mobility and mobility of residences in Langana, Vanuatu. *American Anthropologist* **87**, pp. 56-72.

Rowntree, L. B. and M. W. Conkey. 1980. Symbolism and the cultural landscape. *Annals of the Association of American Geographers* **70**, pp. 459-74.

Sapir, E. 1917. Do we need a superorganic?. *American Anthropologist* **19**, pp. 441-7.

Scheler, M. 1961. *Man's place in nature*. Boston: Beacon Press.

Steward, J. H. 1955. *Theory of culture change*. Urbana, Illinois: University of Illinois Press.

Stocking, G. W., Jr 1966. Franz Boas and the culture concept in historical perspective. *American Anthropologist* **68**, pp. 867-82.

Strong, W. D. 1953. Historical approach in anthropology. In *Anthropology today*, A. L. Kroeber (ed.) pp. 386-400. Chicago: University of Chicago Press.

Thomason, B. C. 1982. *Making sense of reification*. Atlantic Highlands, NJ: Humanities Press.

Tuan, Y.-F. 1973. Ambiguity in attitudes toward environment. *Annals of the Association of American Geographers* **63**, pp. 411-23.

Tuan, Y.-F. 1976. Humanistic geography. *Annals of the Association of American Geographers* **66**, pp. 266-76.

Tuan, Y.-F. 1977. *Space and place: the perspective of experience*. Minneapolis: University of Minnesota Press.

Tuan, Y.-F. 1979. Thought and landscape: the eye and the mind's eye. In *The interpretation of ordinary landscapes*. D. W. Meining (ed.), pp. 89–102. New York: Oxford University Press.

Tuan, Y.-F. 1984. In place, out of place. In *Place: experience and symbol*, M. Richardson (ed.) *Geoscience and Man* **24**: pp. 3-10.

Turner, V. 1982. *From ritual to theatre*. New York: Performing Arts Journal Publications.

Turner, V. 1985. *On the edge of the Bush: anthropology as experience*. Tucson: University of Arizona Press.

Tylor, E. B. 1958. *Primitive culture*. 2 vols. New York: Harper Torchbook.

Varenne, H. 1978. Culture as rhetoric. *American ethnologist* **5**, pp. 635-50.

Vidal de la Blache, P. 1926. *Principles of geography*. New York: Henry Holt.

White, L. A. 1949. *The science of culture*. New York: Farrar, Straus & Cudahy.

White, L. A. 1959. The concept of culture. *American Anthropologist* **61**, pp. 227-52.

Wilson, E. O. 1975. *Sociobiology*. Cambridge: Harvard University Press.

Wissler, C. 1923. *Man and culture*. New York: Thomas Y. Crowell.

Wissler, C. 1957. *The American Indian*. Gloucester, MA: Peter Smith.

10

The language and significance of place in Latin America

DAVID J. ROBINSON

Introduction

To those who are Latin Americans, who share part of that region's culture, who know at least the particuliarities of some of its places, my attempt to generalize about place within the vast continent may appear as arrogant as it is superficial. In my defense I can only say that the observations, interpretations, and opinions which follow come from more than two decades of examining a multitude of Latin American landscapes, of talking to a wide variety of people, of living in a diversity of places for various lengths of time, and of reading relatively widely of past places and events.

In this chapter what I hope to be able to do is no more than exemplify a variety of processes concerning what we have agreed to call "place" within Latin America. I shall attempt to demonstrate the power of place not by recourse to theoretical arguments, but rather by the persuasive evidence of culturally mediated social behavior in specific geographic settings (White 1981). Patterns of behavior, as well as their physical settings, will be emphasized, as will the vital factors of scale and temporal periodicity. I shall argue that place is synergistic: it is created and it creates; that place is constructed, destructed, and transformed by individuals, and/or higher level corporate groups within specific cultural contexts (Robinson 1979a, pp. 22–4). It will be noted, I hope, that I have an uneasy feeling that through time the significance of place may not only have varied but, like culture itself, have been so transformed as to render the usefulness of the general term debatable. Just as a reduction in intimate interpersonal relationships may be one of the costs of the formation of complex societies, so too the restricted confines of small places have usually had to give way to larger geographical entities, most notably the region, nation, state and empire. It is perhaps too easy to romanticize the small (and simple) as beautiful, to characterize the large (and complex) as inhuman and ugly, and yet forget that the modern world in which we live, and from which viewpoint we can comfortably offer reflective criticism, is itself a victory over the confines of a certain type of place (Tuan 1974, 1975, Sack 1980, 1986). What I shall argue is that in each and every historical period there have been advantages and disadvantages of various scales of behavior; if place is significant for social identity, as I think it always is, then we shall see that as identities have to be formed or modified, so too places have to change, from private to public, from narrow valleys to vast regions, from the informal to formal, from community to wider society – and all this normally under the control of an elite minority which establishes the parameters of socio-economic, political, and hence geographical change.

My evidence will be taken from several centuries that have witnessed significant changes in Latin American place, and I shall skip with bold impunity from Caribbea to Araucania in my search for data. I shall also try to invoke and interpret place from as many perspectives and with as many senses as is possible in the space alloted (Robinson, 1969).

The language of place

The word and concept of "place" in the title of this chapter would pose problems for those wishing to translate it into the principal languages of Latin America. In the case of Spanish (and Portuguese equivalents are very similar) it is evident that a variety of alternative terms confront us (see Table 1), each one with significantly different nuances of meaning. Indeed, it is perhaps worth noting, as Mead has done in relation to Finnish, that it is often only when we have to work with a language other than our native tongue that the culturally and temporally specific meanings ascribed to terms become apparent (Mead 1954). Though the majority of the terms in Table 1 derive from Latin roots, the evolution of their meanings, first within Spain, and later in the New World, have given each a ·distinctive meaning. Though in many parts of Latin America *sitio* has come to mean no more than place as site, in post 16th-century Cuba it was a term used to describe rural places that were being farmed (Friederici 1960). *Lugar* too has been used for centuries not only as place in a geographic sense, but also in the alternative English meaning of place as rank or order; to be out of place is to be *fuera de lugar*. Yet to describe somebody as a *lugareño* immediately connotes rusticity. We may also note the occurence of *país* (country) and *patria* (native land), both of which become important ingredients in the identification of national status after the turn of the 19th century. *País*

Table 10.1 Some place terms in Spanish

Place	Populated place
lugar, sitio, situación, localidad, local, terreno, tierra, parte, paraje término, andurrial, ámbito, ambiente, país, región terruño, plaza.	pueblo, pago, población, poblado comarca, nación, país, territorio, patria, ciudad, barrio, distrito

also provides us with the root of the word that Spanish Americans use to denote landscape (*paisaje*). In Spanish *paisaje* has perhaps a much stronger relationship to open countryside (with synonyms of *vista* and *panorama*), than does the term landscape as it has evolved in English usage.

If one is to speak of place as a socially defined area to which persons have a feeling of attachment, then in Spanish the context of the place to be defined is of paramount importance; residents of a major metropolis would probably speak of their *barrio*, or *vecindad* (neighborhood), but if they were newcomers to the city they would still recall their *tierra*, *patria chica*, or *pueblo*. Social classes also appear to have distinctive views of place. Reina has elegantly demonstrated that the significance of the *plaza*, that most important central place in Spanish American cities, varies with the intensity of its use by the *gente decente* and the lower classes (Reina 1973, pp. 76-83). There is also evidence to show that for communities that are in the process of formation the small-scale political unit (*distrito*) may serve as a temporary place identifier. Unfortunately to date few studies have been undertaken to identify the key steps in the temporal process by which urban residents create or become attached to places, and then articulate this place-attachment by means of place recognition. And this in spite of the dozens of analyses of shanty towns and other urban places (Uzzell 1974). What is clear, however, is that the varied density of the socio-economic webs, the complexities of relational associations, be they of mutual trust (*confianza*), of kin and god-kinship (*parentesco* and *compadrazgo*), or of the many other forms by which individuals are integrated within society, all speak of modes of action which are mobilized when occasion or utility demand (Kemper, 1977, Altamirano, 1984, Oliver-Smith, 1986). It is for that reason that our definition of place must remain a flexible one, at least within Latin America, for the contingencies of social action may require what Leeds has called either local or supra-local responses (Leeds 1973).

One fact is clear: a generic term for place as it has become to be used in our discussions in English appears to be singularly absent from Spanish and Portuguese vocabularies of the 16th through 20th centuries.

Even more challenging, is any attempt to understand Amerindian language terms used to denote place. In Quechua, from the northern limits of its use in present day Ecuador, to its southern extensions into Santiago del Estero, Argentina, there exists the term *llaccta* (or *llacctay* in its possesive form) meaning "land of", "my country", or "my soil" (Lira 1944, Ebbing, 1965). This is probably as close a synonym that can be found to represent what we are calling place. In many aboriginal languages such as Nahuatl, Quiché, Aymara, and Ixil, extensive use is made of locative suffixes to attach special meaning to specific geographical places.(Lockhart 1985). For example in Nahuatl the suffixes -*co* and -*tlan* represent what in English would be rendered as "by or near the place of". Generic terms for place are notably absent, which in itself may be an indication that localism was so pervasive within the culture that to speak of generic place was as unnecessary as it was unthinkable (Karrttunen 1983). We must remember that our "place" is a cultural concept whose origins, evolution and use, even within the confines of academic disciplines, have only recently begun to be examined in any detail (Relph 1976, Tuan 1977, 1984, Buttimer & Seamon, 1980).

There also exists yet another means by which we may assess the power of place through its language in Latin America, and that is through the palimpsest of placenames. Toponymic analysis, though still in relative infancy in Latin America when compared to Europe, has provided some of the best evidence to date of the significance of differentiated place(s) in the New World (Raymond 1952, Dykerhoff 1984). For Mexico Moreno Toscano has demonstrated the changing patterns of ranching places (*estancias*) from the 16th through the 19th century (Moreno Toscano 1969); for central New Spain at conquest it is now clear that what became *pagos* (small provinces) for the Spanish had long been *itocayoçanes*, and that the Spanish minor civil jurisdictions were cut out of Nahua *altepetl* (Lockhart 1976, 1985). What place name analysis allows one to understand is both the temporal depth of socially defined places, and the penetration of new name elements. The Spanish adopted and rapidly diffused from the Caribbean island of Española "sabana", a place descriptor of Arawak origin; the Quechua term for a flat grassy place (*pampa* or *bamba*) was likewise included in the lexicon of colonialism and later became the designation of a region in Argentina far from Quechua influence (Friederici 1960, pp. 472-3, 561-2). Gentilic toponyms likewise tell a complex story of culture conflict and diffusion from localities to regions and in some cases imperial levels. How can one explain the sudden rash of Tlaxcalan placenames in northern Mexico without understanding the Aztec colonial policies of the 15th century? Or the Germanic placenames of southern Chile, Brazil and Paraguay without knowledge of the 19th century streams of immigrants? Where settlers or conquerors have gone, normally they have taken their distinctive names with them, and this allows us to reconstruct both the chronology and geography of cultural change (Weibel 1948, Holmer 1960). We should heed Todorov's warning that nomination is often the first step in taking possesion (Todorov 1982, p. 27)

Names have, of course, like places, symbolic as well as purely descriptive content. Thus when regimes fall, empires collapse, or one local elite replaces another, very often a name-changing process is initiated in particular places. In Latin America a veritable national historical geography can often be read from the streetnames of its cities (Reina 1973, p. 67).

We should also recall that whichever island Columbus first landed on in the Caribbean (Judge & Stanfield 1986, pp. 566-99), he was careful to name the sequence of the first five newly discovered places in a rank order which tells us a great deal of the context of his historic enterprise: the Savior (San Salvador); the Virgin Mary (Santa María de la Concepción); the King (Fernandina); the Queen (Isabela); and finally the Royal Prince (Juana) (Todorov 1982, p. 27). The very names of the civil jurisdictions – New Granada, New Galicia, The River of Silver, Little Venice – all speak of the symbolic transfer of sentiment and the imagery of colonial hopes.

The language of place in Latin America thus constantly allows us to interpret not only the mental images and behavior of the various cultural groups, but also provides us with a valuable tool for understanding better many of the physical artifacts that litter the landscape in such profusion. Alas, our Latin American toponymic studies are few, our etymological dictionaries (especially in native languages) still modest in comparison with other culture areas, and there appears to be little interest or expertise in the analysis of names. If, as I would argue, there can be no places without names, then perhaps someone will soon turn to what may be one of the most important research avenues yet to be explored?

Changing places in Latin America

Our next task is to examine a small and select portion of the evidence that demonstrates the changing significance of place within Latin America's cultural evolution during its four principal phases: pre-Hispanic; colonial; republican, and modern. Each will provide us with illustrations of the extent to which social and cultural relationships created, used and valued place, and how cultural complexity can be examined via the contrasting use of place.

Ñawpaq pachakunapi: pre-Hispanic people places

In Latin America it would probably be difficult to find a closer relationship between nature and culture, and behavior and place, than is to be found, albeit still inadequately analyzed, in the pre-Hispanic period. Wherever one looks the aboriginal families, communities, and even proto-states and empires, appear to not only have viewed place as an integral constituent of their culture, but also to have experienced great difficulty in distinguishing between people and place, and social activities and space. If we take as our example the two highest cultures present in the New World at contact by Hispanic man (the Aztecs and Incas), leaving aside the literally hundreds of other cultural groups, we see that the basic units of their society were called *calpulli* and *ayllu*, respectively (Carrasco 1961, 1972, Gibson 1964, Leon Portilla 1984, Murra 1984). Both of these terms are best translated into English as "kin-territories", that is a combination of social relations (with particular emphasis too on lineage and ancestry) within a spatial realm. These were, in fact, people places; or conversely place people. Rules of membership within these socio-spatial artifacts, particularly articulated through kinship and marriage, appear to have been of great significance in limiting their geographic size. Yet another of their features was the emphasis within them of harmonic relationships: bilateralism and complementarity lay embedded in their structure. In Mesoamerica the *calpulli* units often boasted two gods (often the duality of gender, or elsewhere of counterposed forces). In Andean societies the dual moities of *hatun* and *hurin* (the lower and the upper sections of social space) equally symbolized shared power and complementary activities.

If we examine a crude cartographic representation of an ideal-type Andean village we can interpret these patterns with perhaps more clarity (Fig. 10.1) (Isabell 1978, pp. 57-8). The *ayllu* itself is divided into two parts with

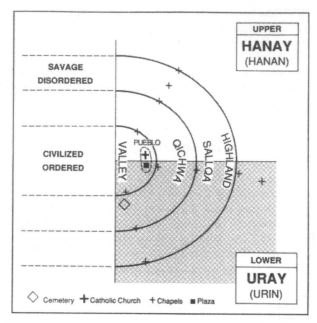

Figure 10.1   The generalised structure of an Andean *ayllu*.

the bipartite village straddling their boundaries. The village itself mediates between the high ground of the puna (the *sallqa*) and the valley floor (the *mayopatan*). Radiating out from the moity core are lines of chapels, each once containing a primitive idol (now a cross), the location of the most significant ones being fixed at the limits of the cultural–ecological zones. These shrine/chapel sites become temporary places of fertility and harvest rituals within the seasonal cycle of the community.

Ecologically the *ayllu* combines both upland pasture and footplow land, an efficient and conservationist vertical integration that Murra has dated from the 15th century (Murra 1975), and which socially demarcates the place of the "savage men" (*sallqaruna*) of the high puna, from their cultured counterparts in the valley bottom. The power base of this ayllu place normally resides in the village, where the *varayoq* (staff bearers) maintain control and mediate disputes on a rotational basis that provides no single individual or group the opportunity to permanently sieze power. To serve as an Andean leader is thus to make a dramatic, short-term, economic sacrifice for the benefits of prestige and a permanent place in the community's social fabric. In the highest reaches of the puna dwell the deities, the *Wamanis*, guarding the lakes and peaks, and to whom ritual payments must be made when the earth is "open" to such offerings.

Our Andean microcosmic place, therefore, reflects a synchronization of resource use, a mediation of social conflict by way of dual prestige hierarchies, and presents a namescape which almost defies attempts to list let alone map the hundreds of identified places on the ground (Platt 1978).

Sacred and profane here still have an ecological and cultural meaning that most Europeans find difficult to describe, but which they recognize to have been integral ingredients of *ayllu* and *calpulli* alike.

But of course some will be already asking how these microcosms related to each other, and what was the potential for larger scale developments. To partially answer that question we may turn to the evidence of Inca imperial structures, viewed through the perceptive eyes and skillful pen of a 17th century alleged descendent of one of the last Incas (Adorno 1986, pp. 80-120, Pease 1985). Huamán Poma de Ayala provides us with one of the most detailed images of the power of place for Inca culture at large. A careful examination of his mapamundi (Fig. 10.2) reveals that in spite of its European imitation of form, its content is rich in the symbolism of place. Its central settlement (where of course one would expect to find Jerusalem or Rome after a century of Spanish rule) is Cusco, the Andean pivot of the four quarters. Indeed, if we clarify the details of the indistinct lines on his map we can distinguish those four quarters by name: the four *suyos*. The whole is no less than a symbolic representation of Tawantinsuyo, a quadripartite image of the Andean universe.

More interesting details can now be identified. The four quarters themselves can be seen to be divided into two pairs, two forming an upper, and two a lower sector: the model of empire and universe can now be seen to contain the same *hanan/hurin* that we noted earlier in the *ayllu*. This is an ordering of places and spaces that Zuidema and others have identified as a basic structure in Andean culture as far back as our evidence permits us to look (Zuidema 1964, Lopez-Baralt 1979, pp. 83-116, Harrison 1982). Other ethnohistorians argue that upper and right may be equated with maleness, which itself counterpoints the complementarity of lower/left femaleness. Lest it be thought that Huamán Poma's representation is no more than a theoretic construct, it should be mentioned that archaeologists are finding in ever-increasing quantities, information that demonstrates that the limits of the four *suyos* (the *ceque* lines) were not only often demarcated on the ground, but that local community structuring was also affected by their presence.

Thanks to recent ethnohistoric analysis we also now know much more of the significance of the four quarters in the power relationships among the various competing groups. To achieve access to the most central place (Cusco) was the aim of all contenders to Inca authority. From the temples on the open-sided great square in Cusco, beneath the belly of the symbolic puma that shaped the form of the very city itself, from between the metaphorical legs of the gods a vast empire could be controlled (Chávez Ballón 1970, pp. 1-14)

Any attempt to summarize the place elements of the pre-Hispanic world necessarily should stress not only the interdependence of cultural form, social behavior and place, but also the manner in which places small and large were nested hierarchically and, perhaps more important for our immediate concerns, paralleled in the positions of power held by the Inca elite. To be posted to Quito or central Chile was the worst fate that could await an official of the empire.

Mapamundi of Tawantinsuyu

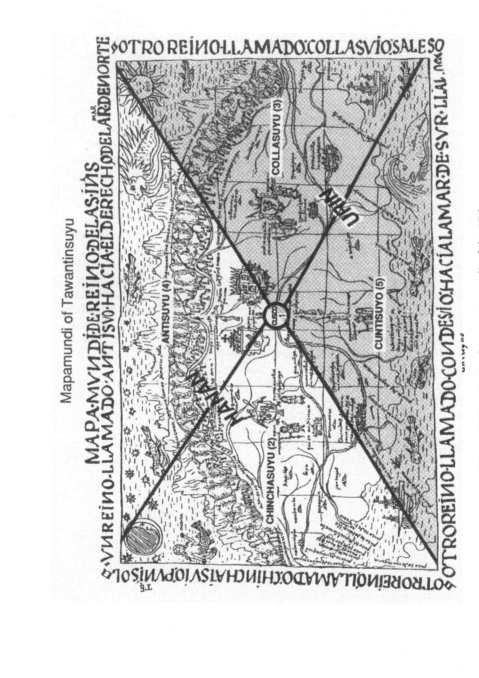

Figure 10.2 An Andean mapamundi of the 17th century.

Equally important was the manner in which in the period immediately prior to Hispanic contact these macro-structures were undergoing stress and strain as the supra-local power found itself challenged both from within its own increasingly segmented and hierarchical society, and from external challenges from non-sedentary groups (Stern 1982, pp. 51-80, Spalding 1984)

Putting colonialism in place

Our second stage, that of Hispanic colonialism, provides us with a rich diversity of behavior acted out now on a quite deliberate stage: that of imperial powers each bent on establishing their own cultural identity. In many instances that could best be accomplished by means of policies designed to undermine, and overthrow the power base of the aboriginal cultures, as well as a persistent process of cultural restructuring (Wachtel 1971). At the heart of the latter lay a new set of place rules and regulations with which both Spain and Portugal (to a lesser extent) attempted to repress the Indian population into following. The aboriginal people (at least those who did not die from the accidental onslaught of epidemic disease) had to be firmly put in their new place: and that place was decidely under White political control, in the deep shadows of racial segregation, and physically separate from the newcomers (Lockhart & Schwartz 1983).

The Hispanics brought with them to the New World a set of institutions, a complex of cultural artifacts, and patterns of behavior that immediately distinguished them from their antecedents. The Catholic faith, the extended family, the urban system, the desire for and conspicuous enjoyment of material wealth – all these and more had serious consequences for placemaking and place-changing in Latin America.

Perhaps the most symbolic of the new places created in Spanish America were the scores of urban places that were established throughout the length of the rapidly expanding empire. The checkerboard street system, the central place (the *plaza*), the formal and functional differentiation between central and peripheral all bespoke a new cultural order (Nuttall 1921–22, Borah 1972, pp. 35-54). The most humble of settlements, some which we would perhaps hardly wish to designate urban, included these vital elements. At their heart lay the plaza, that open space upon which had been set the twin symbols of Spanish imperial power, the sword and the cross, that was to function for centuries as *the* place in which to socialize, via public meetings, or the courting ceremony of the *paseo*; the place in which to conduct business, be it via the market stalls or the surrounding covered archways; the place to locate one's house close by; the place to behave in the proper manner of a *vecino*, a proud and registered resident of the town (Tagaki 1970, Gade 1974, Robertson 1978).

On the edge of the plaza were sited physical representations of imperial power: the church, the town council, the viceregal office, etc. Often these buildings presented their significance to society at large by not only their location, but also their elaborately adorned two stories, often no more than facades that hid the more mundane reality of the town's aspect. The toll of the bell and the firing of the cannon, remind one of the *campanilismo* of Mediterranean antecedents, where to belong to the settlement one had literally to be within earshot (Pitt-Rivers 1961, p. 30).

These urban places were legally not for residential use by Indians: to protect the latter from possible cultural contamination from the vices of the Hispanics, they were provided with special settlements at discreet distances from the White towns. The Indian *pueblos* replicated the form of the Hispanic towns, and Indian leaders were alloted similar functions, albeit of a more restricted kind, to those of the *gente decente* of Hispanic origin (Gibson 1964)

This "civilizing by settlement" promoted major campaigns to relocate the dispersed Indian population. A major goal in resettling them in nucleated villages (the famous *congregaciones* or *reducciones*) was to allow better political and adminstrative control, and to facilitate the required conversion of the heathens to the new Catholic faith (Cline 1949, Fals Borda 1956, Málaga Medina 1975, Lovell 1985). It is important to note that the move downslope of the aboriginal population, besides representing a new colonial phase in agriculture on the plowable bottomlands, also meant a distancing of them from their mountain gods. It is evident too that in the new villages, recreated moity structures were fatally penetrated by the intrusive central place which unbalanced the duality of aboriginal places.

Where conditions made civil control difficult a set of new places was established –the missions of the regular orders (Ricard 1933, Mörner 1953, Specker 1953, Robinson, 1975, pp. 63-76). Hundreds of abandoned and ruined churches, some recently restored as national monuments, litter the landscape of contemporary Latin America.

Within these colonial urban places, the primary foci of imperial power, one can recognize yet another scale of place-making: that of the house and home. The micro-ecology of the colonial town reflected the cultural norms of the intrusive Hispanics. The high-walled blank periphery of the colonial house demarcated a private world beyond the sounds and (normally bad) smells of the city. In the best of cases, access to these private places was via an imposing doorway, that spoke of the status and prestige of its owner. The fortunate few might even boast a coat of arms to match their title of "Don" . Passing through the entrance hall, beyond lay the designated patios, each of which in ranked order provided a focus of racially and functionally differentiated activities. The further back one moved from the front door, the closer one approached the servants' and slave quarters (Torre Revello 1945, Gasparini 1962, Guarda 1978, Robinson 1979).

I would like to suggest a symbolic and functional parallel beween the urban-scale places (ceremonial arch, processional ways, plazas), and those of the residential units (doorway, passages, patios). In both public and private arenas one was informed of proper behavior, dress, language, and the like by one's relative location (Hoberman 1986, pp. 325-8)

On special occasions the social codes and power relations within the colonies were temporarily suspended in ritual inversions which permitted the humble to parody the powerful, the poor to ridicule the rich, the Blacks and Indians to literally put on the masks of the Whites and all to enjoy the fleeting freedom of equality unabashedly assisted by the intake of large amounts of liquor. By means of public fiestas and carnivals people could, so to speak, change social places. It provided an excellent means of reducing the many social tensions.

Public processions and seating arrangements in the church and on stands erected in the plaza also provided occasions for not only a display of social position, one's power demonstrated by one's proximity to the altar or portable icon, but also frequently provoked fights and resulting law suits. The terms *"preeminencia de asiento"* (privilege of seating) and *"orden del desfile"* (rank procession order), invoked in the many legal cases speak clearly of a society that attached great value to place (Colombia, 1796). Death provided yet another opportunity for one's descendants to memorialize one's place in society by a well-sited and suitably impressive tomb. The mausoleums and niches of the cemetery cities in most Latin American countries speak eloquently of this mode of posthumous privilege (Reina 1973, p. 373).

Yet we must remember too that the ideal types of Hispanic city and Indian village rarely survived in any way intact into the 17th century. Racial mixing, economic development, and population migration soon appears to have blurred the theoretical sharp divisions between distinctive places. Around, and often within, cities there grew the Indian sectors (*barrios*), often provided with their own parish facilities. Some members of the elite soon realized that there were profits to be made from renting portions of their centrally located houses to artisans and traders (Góngora 1975, pp. 421-48, Moreno Toscano 1978, Ramos 1979).

The city residents were also elements in a larger spatial web of economic interests, of social obligations, and political power by dint of family connections. If being a "proper" person was to be somebody from a known place and of distinguished ancestry, then so much more important were persons who had access to resources over a wide geographic span. By the 18th century the notable families of Latin America could boast of networks that connected mines to haciendas, slave estates to urban residences, and monasteries to the imperial court. The full significance of regions as complexes of interrelated extended families is only just becoming evident (Kicza 1983, Balmori *et al.* 1984, Kuznesof & Oppenheimer 1985).

And yet what do we know of the identity of places within and without these evolving cities ? The answer unfortunately is relatively little. It is clear that urban neighborhoods obtain names as time progresses, but as yet it is difficult to ascertain the role of residents in this process (Borah 1984, pp. 535-54). It is evident that within two generations of residence Hispanics begin to speak with exaggerated pride of their newly created urban-based *patrias chicas*. Inter urban/regional rivalry, perhaps a useful indicator of identity, by the 18th century had reached proportions which began to alarm the imperial courts. And there is also another measure by which we may estimate the significance of place for colonialists of both urban and rural persuasion. We can see in the terms used to describe migrants, both temporary and permanent, who in Hispanic eyes belonged to no fixed place (the *vagos*, *vagabundos*, *huidos*), the threat they posed to the stability of the social order (Robinson, in press). If a person was not known in a locality, was not of a place, how could one judge his ethnicity, his morality, and his worth to the community? Transients, be they artisans searching out a livelihood, Indians escaping harsh taxation, or the ever dubious and growing mestizo class – all were suspect. Indeed legislation was frequently though unsuccessfully enacted to put these people back in their proper places (Farriss 1978).

But perhaps the best metric of attachment to place in colonial Latin America is afforded by those who lost the privilege of living there. It is in the anguished laments of the exiled Jesuits of the 18th century that we can establish the first continent-wide identity with Latin America as a place. Perhaps this Americanization or creolization was a consequence of what Lynch has described as the second conquest of America ? (Lynch 1973, p. 7 *et seq.*) Only when the criollos had quickly learned to hate the peninsular newcomers (the *gachupines*) who had arrived with the Bourbon reforms, did an appreciation of their special place – America – begin to emerge. The conviction that Americans were not Spaniards, but rather Venezuelans, Mexicans, Chileans, and Peruvians, could perhaps only have emerged as a result of the long centuries of colonial development. In making their own places, in creating their own identity, the Hispanics had finally to reject their origins in the Old World.

The exiled creole Jesuits became the literary precursors of American nationalism, and in no small measure it was a literature of nostalgia. Manuel Lacunza "imagined himself eating his favorite Chilean dishes, while Juan Ignacio Molina thirsted for the water of the cordillera. The Mexican Juan Luis Maneiro implored the Spanish king to allow him to die 'en patria suelo'" (Gonzalez, 1948, p. 158).

Since a precondition for love of one's own place (patriotism) was knowledge, the pamphlets and newspapers of the late 18th century burgeon with facts of the geography, the resources, and the potential of this new promised land. Uninhibited Americanism was flaunted; "*la patria*", "*nuestra nación*", "*nuestra América*", "*nosotros los Americanos*" (Burrus 1954, Gravales 1961, Vial Correa 1966). One can understand the preoccupation of the imperial authorities: love of American places appeared dangerously to have eroded political allegiance. Soon many Americans were to be called upon to pay the ultimate price for their patriotism.

Revolution and republicanism: places new and old

If the struggle to end colonialism proved difficult for Latin Americans, no less arduous was their quest for a set of new identities (Lafaye 1976). It was easier to be free of Spain and Portugal than it was to become Brazilians, Argentinians, and Guatemalans (Stabb 1967, Gossen 1974, Hawkins 1984, pp. 67-87). The challenge now was to forge new socio-political alliances, to instill within the liberated population a care for, and pride in, their newly won place in the world. And as in almost all such circumstances the most difficult issue was to decide just how much of the old to reject, and how much of the new to afford. The task of examining place within Latin America now became much more difficult since each and every one of the new republican nation states addressed this central question in a distinctive manner. Here, one can hope to do no more than illustrate tendencies in processes that diffused thoughout the continent. One should also stress that the pace and direction of change was almost never synchronized.

Statehood first demanded the repression of regional rivalries (Seckinger 1984). The authors of this new macro-regional structure, and artifice of political and adminstrative control based on western European models, rapidly set about manipulating and mediating regional power in the alleged interests of the unwitting citizens of the new republics. In Argentina Rosas' police state was soon using *mazorcas* to "clean up" the new republic, red-pepper enemas serving the same purpose as the castor oil later to be used by the Italian fascists (Arciniegas 1966, p. 358, Lynch 1981).

Ubiquitously and simultaneously a single capital place now began to set itself apart from the rest, a coastal site normally being a *sine qua non* since trade and overseas connections were increasingly to orient the new "national" life. Littorals now confronted interiors, and several colonial peripheries attained central positions (Humphreys, 1957, 1969, Scobie 1964, 1974, Eidt 1971.) In both unitary and federal nation states the dominant center now began slowly but surely to exact its toll on its internal regional dependencies. For those in the newly advantaged central place, the national illusion was fostered and reinforced by symbols of centrality and cohesion: the National Palace, the National Treasury, national laws and national maps. Yet it is clear that in many respects such "nationalism" was little more than a new myth. Just as the new elite adopted new poses, ideologies, and policies to ensure their own stability, so too their states remained little more than images viewed from on high or from without until the authoritarian regimes of the late 19th and early 20th centuries, when weak state apparatus were strengthened, and the population at large was once more told where they belonged, and how to behave (Brading 1973). In Mexico, dictator Porfírio Diaz could refer to his legislators as "my herd of tame horses" (Simpson 1952, p. 263). Mass education became not merely a modernizing medium, but a necessary tool of national indoctrination (Campos Harriet 1960, Spalding 1972, Pike 1973, Woll 1982). If the state needed well-defined boundaries (for *uti possidetis* had proven quite unclear!), so too the citizens needed an anthem and a flag . Recognition as a strong state often required a common enemy, and neighbors served that purpose well (Clissold & Hennessy 1968). Cultural continentalism (*panamericanismo*) soon foundered upon the rivalries of competition for leading places in the new economic and political order (Hilton 1969, Karnes 1983).

In urban areas the built environment was modified to celebrate the triumph of political independence: boulevards and processional ways opened vistas to the juxtaposed artifacts of the nation state. The houses of congress confronted the presidential palace, the latter often sited symbolically in or on top of the old viceregal administrative complex. (As in the 16th century the best place for one's enemy was under one's feet.) (Scobie 1974.) Never far away were the barracks of the military, who unfortunately soon learned the opportunities afforded them as guarantors of the constitution. The statuary of the plazas reflected the new national heroes: San Martín; Admiral Brown; Bolívar; O'Higgins; Santander.

Beyond the symbolic also began to appear the new functions of national and regional metropoles: a business district with banks and branches of the leading commercial houses of Europe; new residential quarters for the rich elite who emulated the latest architectural styles that they had seen on their frequent vacations to Europe ; and new crowded slum districts and tenements (Morse 1978). With the British ever ready to loan capital to the new national elite, slowly but surely new technologies could be imported. Public services were provided for the fortunate few; tramways and railroads extended the reach of the expanding cities, allowing land speculation and a gradual new patterning of urban places (Scobie 1972, Sargent 1974, Goodwin 1977). The elite suburbs, constructed often from the profits made from the sale or renting of centrally located properties, contrasted ever more sharply from the poor central districts (Hardoy 1955, Benitez 1965).

In several countries the changing physical fabric of the city was paralleled by a new ethnic mix. A virtual tide of immigrants rapidly sorted themselves out into ethnic enclaves. British, Jewish, German and Italian settlers now provided cultural identities for the new sections of town (Solberg 1970, Bourde 1974, Newton 1977, Szuchman 1977, Baily 1980, 1983, Sopher 1980, Whiteford 1981).

And each ethnic community provided the necessities for its recreational as well as occupational needs. The British with their polo and cricket clubs; the Italians building bocche pitches; and the creoles, not to be outdone, creating what they imagined to be the proper pastimes for decent men and women – boating with the Regatta Club on Lake Xoximilco, running with the hounds in Santiago Chile (Mayo 1987), a day at the races with the Jockey club in Buenos Aires – and everywhere the coffee shop, the smoking parlor, the tea room, and the discreet gambling saloon (Bossio 1968, Scobie 1974).

The old colonial city was undergoing dramatic changes. The workplace was no longer coterminous with the home; the new houses now modeled themselves on European fashions and relocated the internal patio as a green skirt around the building, normally protected from improper public use by an imposing and imported wrought iron fence. These new *palacios* spoke of wealth and power. One now could identify people not only by the location and style of their residence, but also by the language they spoke (or the accent they affected), their dress, their preferred food and drinks, even in some cases the very way they walked and gestured; and always by their social connections (Needell 1983, Caldeira, 1986). Places in urban republican Latin America proliferated rapidly.

In rural areas new landscapes were also being created: the technological innovations of Europe and the USA allowed millions of acres to be fenced, and plowed for the first time. New crops, new agricultural laborers, and new markets stimulated economic growth and regional identity. Elsewhere the mining of first nitrates gold and guano, and later copper, tin, iron, and petroleum gave rise to new wealth which was only occasionally siphoned back to the urban centers, the vast majority of the *nouveau riche* preferring to not just mimic the Europeans but join them. One should not forget that to meet the truly rich Latin Americans, by 1900 one would have had to visit the hotels of London, Paris, and especially the French Riviera.

To the colonial Chilean Norte Chico was thus added the mining Norte Grande, and the German colonized forest south (Berninger 1929, Butland 1957, Pederson 1966, Blancpain 1974, Vayssiere 1980, O'Brian 1982). Argentina witnessed the creation of its new pampas region, and in the deep south strange Welsh folk struggled to establish their Cwm Hafryd in Chubut (Bowen 1966, Slatta 1983). In central Venezuela Germans built their own tropical version of the Black Forest, and in Venezuelan Guayana British and Trinidadians so successfully mined the Caratal goldfields that Venezuela feared for its very sovereignty (Robinson 1973).

In Brazil, Venezuela, and Costa Rica coffee lands prospered, and German colonization proceeded apace (Franca 1956, Hall 1976, Bergquist 1978, Holloway 1980, Roseberry 1983). Even remote Amazonia felt the rubber-clad hand of development (Weinstein 1983), the Patagonian steppes attracted the attention of Welsh and Scottish colonists (Rey Balmaceda 1961, Williams 1964, Williams 1976), and the pampas their Irish counterparts (Korol & Sabato 1979, 1981).

All of these new places, and transformed older places, symbolized their new identities with names. El Dorado, that gilded man-place so diligently searched for by the Spanish, became a label for dozens of mineral strikes and probably thousands of farmsteads. Llaneros of the Colombian and Venezuelan plains heard of their *pampa* counterparts, the gauchos (Nichols 1968, Rausch 1984); the Peruvian *mistis* and *cholos* now became *ayacuchanos, ancashinos,* and *cusqueños.* Hardworking *paulistas* and *antioqueños* compared themselves favorably with the *yanquis* of the north, and belittled the life styles and pretences of the *cariocas* and *bogotanos* (Morse 1958, Parsons 1968, Dean 1969). New Chicagos, New Filadelfias, New Californias, New Providencias – the new places began to clutter the carefully constructed cartographic records.

And yet we have also to remember that the tide of immigrants was not always warmly welcomed. Many in Argentina, Chile, and Venezuela viewed them as suspect citizens; some asked just whose country it was when whole sectors of Buenos Aires spoke German and English. Antisemitism also raised its ugly head. Clearly an open-door policy on immigration, basically to meet the demands for cheap labor, had serious potential consequences. "To govern is to populate" was the dictum of the day, but some asked just who would be governing whom? By the turn of the century the Chilean newspapers could speak of immigrants as "dirtier than the dogs of Constantinople", and "waves of human scum thrown upon our beaches by other countries" (Solberg 1970, p. 71). Ramos Mejía (1899, pp. 255-8) described the newly arrived Argentine immigrant as "a gross person, one of those low beings that future scientists will study with curiosity in order to establish the linkage of the successive types of our evolution. With their cheap, sensual tastes and their love of bright colors, raucous music and gaudy clothing, they are simply inferior".

Just like the Indians of the Andes and Mesoamerica, the new immigrants were to be used for nationalistic purposes only when it was in the interests of those in power.

Making room for modernity

In the headlong rush to create and modernize the new states there were, of necessity, some regions and localities that did not enjoy or suffer even the minimum of progress. There was, after all, never enough money to go around for everyone. One could soon identify the initiation of a new patterning of place: places of plenty and places of neglect, of provincialism now perceived as an unhealthy attachment to traditions that conflicted with the new demands of the state (Love 1980, Wirth 1977, Levine 1978, Demelas 1980, Montoya 1981, Weinstein 1982, Mallon 1983, Gonzales 1985). Ethnic cleavages combined with physiographic provinces to delimit problem places. Those who did not participate in the political process, irrespective of whether they were disallowed by law, or were simply ignorant of the entire matter, soon found that the response of the powerful was at best benign neglect (Schmitt 1969). When, on occasions, the inconvenience of democracy demanded votes, the regional bosses could always be trusted to bribe or threaten the necessary number of peasants (Brading 1980). Political independence of the early 19th century had, after all, done little to change the order of affairs on the agricultural estates.

By the 1880s it had become clear that to advance in anything one had to approach the city, and for the next century first a trickle and later a flood of migrants sacrificed the known inadequacies of their rural and small-town settings for the promises of the city (Bergquist 1986, Hall & Spalding 1986). Twentieth century Latin Americans have usually had to walk towards modernity (Graham & Buarque de Holanda 1971, Castellanos de Sjostrand 1975, Hagerman 1978, Laite 1981).

Occasionally a rash of rural protests might break out, but the possibilities of contagion were easily limited by the relatively efficient combination of police brutality and minimal offers of reform (Benjamin & McNellie 1984). Even in the wake of revolutionary change in Mexico, that reconstructed the agricultural landscape of vast areas, it was not difficult to find those who were willing to sacrifice the potential of radical change for the less risky, and certainly more comfortable, benefits of conformity. Only under the most provocative and repressive of circumstances have Latin Americans opted for armed struggle and open rebellion, and then often in liaison with non-local forces and ideologies (Womack 1969, Mayer 1973, Katz 1976, Ruiz 1976, Joseph 1982, Cockroft 1983, Wasserman 1984). One of the costs of localism may have been the inability to perceive of the possibility of changing the status quo of the state, or the socio-economic and political systems that provide its support.

One thing is certain: in the last 20 years Latin America has witnessed a resurgence of local and regional identities and actions, within both urban and rural realms. In their move to the cities migrants neither severed ties to their homelands, nor assimilated into a homogenously bland urban culture. On the contrary, evidence is available in almost every country to demonstrate the maintenance of social ties and cultural roots (Doughty 1974, Roberts 1974, Collier 1976, Hirobayashi 1986). The regional clubs (and Lima in 1985 had some 1,850) provided a new institutional mechanism that replaced residential propinquity as the *sine qua non* for community identity. The intensity of weekend meetings appears to more than compensate for the lack of day-to-day contact (Altamirano 1984). Their place in the city is thus not a segregated residential zone, but rather a meeting point. And perhaps the same was true even in their Andean valleys ?

When the state was unable to provide these newcomers with shelter and services they did what they had long done in their rural communities – help each other. The millions of *barriada, rancho* and *favela* houses, all completed without the benefits of civil engineers and architects enamoured of concrete and style, speak eloquently of local solutions to local problems. The *faenas* and *turnos* have accomplished what no Ministry of Housing with international aid has even been able or willing to do (Turner 1977, Portes 1979, Conway & Brown 1980, Lobo 1982). Of course in the eyes of those who prefer elegance bought with mortgages, the artifacts of these peripheral communities were objects to be condemned, reformed or eradicated, to be relocated beyond the sight of the more affluent suburbanites (Epstein 1973, Holsten 1986).

Occasionally city fathers would call upon alleged experts to advise them on how to cope with the increasing problems of metropolitan sprawl. Rio de Janeiro provides an illustrative case. Le Corbusier, who was invited to visit Rio in 1929, admitted to being inspired by the sheer beauty of the city's location, so much so that he admitted "to a strong desire, a bit mad perhaps, to attempt a human adventure – the desire to set up a duality, to create 'the affirmation of man' against or with 'the presence of nature'" (Evenson 1973, pp. 52-6). This affirmation, however, was to have taken the form of an immense motor freeway 100 meters high with apartment blocks below, which would be a "poem of geometry..." Fortunately for Rio this poem remained on paper and one can still enjoy the chaotic beauty of Copacabana and Ipanema.

Politicians were also quick to appreciate the potential benefits of strongly held feelings, at the local and regional levels (Oszlak 1981). Most of the great political issues of 20th-century Latin America can be seen to have originated in the context of some specific set of local circumstances (Bushnell 1954, Wortman 1976, Flores-Galindo 1977). The conservative highlanders of Quito soon clashed with the brash liberals of Guayaquil who saw only benefits in a new commercially-based social order (Alaya 1978, Deler 1981)

Aprismo, though initially conceived of as a pan-American alliance, required the stimulus of the appalling work conditions of the sugar estates of northern Peru, and the local military repression to build its base (Klaren 1973, Vanden 1979). Only after deliberate policies of diffusing the message of the party could it later challenge, and finally overcome the prejudices of the military and the majority of the Peruvian electorate. FRENATRACA, the new league of Puno peasants likewise is currently expanding its regional base (Rojas Samanez 1985). In every country the roots of party ideologies can be traced back to localities, to individuals in contexts that demanded reappraisals of social justice and state policies (Magallanes 1973, Schwartzman 1973, Carvalho 1980, Hardoy & Langdon 1982, Anderle 1985, Park 1985).

For the ideological left the city, with its flagrant display of wealth and yanqui capitalism, has provided a notable target for criticism and recruitment of supporters. When the campesinos either rejected or were not permitted to accept the offer of armed revolution, guerrillas and state-defined subversives also moved to the cities: Sandinistas, Fidelistas, Tupamaros, Montoneros, Senderistas – the list is long (Gott 1970, Kohl & Litt 1974, Gillespie 1982).

For many who wished for speedier social change during recent decades their communities and their places became their tombs: in Ayacucho's uplands, in Uchurraccay, in dozens of yet unidentified places *los desaparecidos* have made the ultimate sacrifice for their traditions old and new (Thorndike 1983).

But beyond the local and national scales, Latin America's relative place within the power structure of the continent and the expanding world political economy also slowly began to become established. John Bull's greedy grab at resources in the 19th century was followed in the 20th by Uncle Sam's paternalistic attention. The Latin "children" could be kept happy within the protective custody of Mr Monroe; errant "daughters" could be gently persuaded to behave properly; even angry neighbors could be suffered, although not very gladly, and a reformist sore throat, or a revolutionary ulcer could be taken care of by dollars or, when necessary, by force (Johnson 1980, pp. 116–54, 233). Latin Americans collectively began to be conscious of where they had been placed by their more powerful neighbors to the north. They noted that the "Americans" now were Anglos and not Latins, and they deeply resented the fact that they were viewed as a virtual embarrassment to hemispheric progress. Once formed, stereotypes of people and places have proven difficult to modify.

Yet within Latin America itself the problems of place continue. Attempts at regionalization and the decentralization of economic development have frequently foundered on the rocks of regionalism, of people not wanting to follow the prescriptions of spatial scientists and power brokers (Whitehead 1973, Harris 1983, Delgado Medina 1984). Feeling for place has surprised more than one politician or international development consultant. Unfortunately, we still lack the technical means to include such feelings into planning and policy decision-making, so that conflicts and tensions continue.

There is, however, at least one direction in which place and love of places is making significant positive advances; that is in the rapid development of the tourist industry which has perhaps done more for place preservation than anything else during the last 20 years. Few are the countries that now do not have a ministry that attempts to direct vacationing visitors to a variety of notable places. The quaint, the traditional, the remote, the exotic – these are places that have gained a new lease on life. Even better if they have a photogenic folklore with music and songs (the less intelligible the better), dances, handicrafts, and strange customs. Suddenly the modern world is anxious to visit places, and Latin America is fortunate to be high on the list of preferences (Bryden 1973). Even internal tourism is now a boom industry in many countries for the urbanized population is now being persuaded to "discover" its own country again. Modern transport has made almost any place a possible site or route.

Conclusion

Though this chapter on the significance of place in Latin America has only permitted glimpses of a subject that demands more detailed attention, some conclusions may be drawn. First, there is the ever-present issue of scale. If places are social constructs then how best may we compare say the house or home to the nation state? Perhaps we should be looking to the operational mechanisms that allow people to attach themselves to places of quite different types? At the level of the domestic unit or small village kinship may be of significance in the intensity and frequency of interaction. But at the level of systems of cities only perhaps some families can maintain ties, the remaining population resorting itself amongst the proliferation of other institutions such as clubs, schools, neighborhood stores, etc. Thus perhaps place attachment does not weaken with modernization but rather is transformed into new and subtle guises. One is able to be a person of several places rather than one; and one may be of any one of those places for different durations of time. One has only to compare the life-histories (or geobiographies) of distinctive groups within society to see that one's "place potential" greatly increases as the culture becomes more complex (Hernando 1973).

Second, it is clear from the evidence of Latin American development that each culture has its own particular manner of using and assigning a value to place. But even culture may be too coarse a concept with which to estimate the significance of place. One has only to examine the differential results of place creation by the rich and poor classes of the same culture to see that finer social gradations of analysis may be necessary. Equally important is time: within the same social class place may have a quite distinctive significance from one period to another.

Third, experiential place – those places that one knows personally, through the soles of one's shoes, through suffering in the weather, through loving or hating the people who live there – these types of place I would wish to distinguish from image places – places that one can rarely visit or live in, or experience in the same sense that one can in the previous class, but places for which one might be prepared to fight or die (Tuan, 1975). These places are much more mental constructs in the sense that the information upon which we base our feelings is significantly indirect. One can think of the comments of a Franciscan friar who, in the 16th century, spoke of "The Indies", when it is unlikely that he had ever been outside the Caribbean island of Española. Or the Argentine literati's views of the savagery of the

175

19th century countryside when they had rarely left the comfort of Buenos Aires (Franco 1969, p. 46 *et seq*). Or the willingness of Argentines to send their sons to die for a small island group in the icy waters of the South Atlantic (Coll & Arend 1985, Gamba 1986). This is place consciousness and feeling that is quite different from that of people whose places are of their own creation.

Fourth, the dual meaning of place in English (place as location and as rank or order) fits very well the Latin American patterns of place cognition. The coincidence of spatial centrality and social importance in the colonial period is self-evident. Even more interesting is the fact that the socially important (the elite) could change their geographical places so significantly, and yet maintain their social rank. One should perhaps add to their many other attributes, the skill of maintaining prestige while changing identity.

Fifth there is the persistence of place. How difficult it has been in Latin America to completely eradicate places once constructed or identified. Perhaps social memory is as well served by landscape artifacts as it is by any written text.

And finally there is place as context of control. One can climb the sides of the pyramid of power in the colonial period – on the ecclesiastical side from confessional to church sanctuary, to atrium, to parish, to see, and finally to Rome. There one can intersect via the Patronato Real to civil authority and descend from Council of the Indies through viceroyalty and intendancy to *corregidor* and *alcalde*, each level with its proper place and symbols. It is similar in the modern period when governent agencies plan for the powerless who have to inhabit new places, be they colonization villages in the Amazon, or discreetly located construction workers townships outside of Brasilia (Leff 1982).

Size too may make a difference. After all there comes a confidence with bigness. Who in mighty Brazil really cares what is happening in the backwoods places of Honduras ? Or who in Mexico even knows where the Peruvians and Argentinians are proposing to relocate their capitals to? We have to remember that what to one is sacred to another is profane – desecration and resource utilization are often two sides of the development coin. There is perhaps always a comfort in planning places in which others have to live.

One thing is for sure. If nothing else Latin America provides us with abundant evidence of the persisting significance of place. The only thing that remains is for Latin America to achieve the place that it (or collectively its peoples) insists that it deserves within both the hemisphere and the world. With Washington at the center of the contemporary mapamundi, that would appear to be at best no more than a faint hope. But Latin Americans have learned through the experience of centuries that hope of a future place may be the hallmark of their cultural heritage.

References

Adorno, R. 1986. *Guaman Poma: writing and resistance in colonial Peru.* Austin: University of Texas Press.

Alaya, E. 1978. *Lucha política y origen de los partidos en Ecuador.* Quito: Editorial La Paz.

Altamirano, T. 1984. *Presencia andina en Lima metropolitana: estudio sobre migrantes y clubes de provincianos.* Lima: Pontíficia Universidad Católica.

Anderle, A. 1985. *Los movimientos políticos en el Perú.* Havana: Casa de las Américas.

Arciniegas, G. 1966. Civilization and barbarism, In *Latin America: A Cultural History* pp. 351-77. New York: Barrie & Rockliff.

Baily, S. L. 1980. Marriage patterns and immigrant assimilation in Buenos Aires, 1882-1923, *Hispanic American Historical Review,* **60**, pp. 32-48.

Baily, S. L. 1983. The adjustment of Italian immigrants in Buenos Aires and New York, 1870-1914, *American Historical Review,* **88**, pp. 301-25.

Balmori, D. *et al.* 1984. *Notable Family Networks in Latin America.* Chicago: University of Chicago Press.

Benitez, I. 1965. *"Los Olivos". Barracas al Norte, 1895-1960. Para la antología de los barrios porteños.* Buenos Aires: Oveja Negra.

Benjamin, T. & W. McNellie (eds) 1984. *Other Mexicos: essays on regional Mexican history, 1876-1911.* Albuquerque: New Mexico Press.

Bergquist, C. 1978. *Coffee and conflict in Colombia, 1886-1910.* Durham, NC: Duke University Press.

Bergquist, C. 1986. *Labor in Latin America: comparative essays on Chile, Argentina, Venezuela and Colombia.* Stanford: Stanford University Press.

Berninger, O. 1929. *Wald und öffenes Land in Süd-Chile seit der Spanischen Erobung.* Stuttgart: Böhlau Verlag.

Blancpain, J. 1974. *Les allemands au Chili, 1816-1945.* Cologne: Bölau Verlag.

Borah, W. 1972. European cultural influences in the foundation of the first plan for urban centers that have lasted to our time. In *Urbanización y proceso social en América Latina,* R. P. Schaedel *et al.,* (eds) Lima: IEP.

Borah, W. 1984. Trends in recent studies of colonial Latin American cities, *Hispanic American Historical Review,* **64**, pp. 535-54.

Bossio, J. A. 1968. *Los cafés de Buenos Aires.* Buenos Aires: Paidos.

Bourde, G. 1974. *Urbanisation et immigration en Amérique Latine: Buenos Aires.* Paris: Université de Paris.

Bowen, E. 1966. The Welsh colony in Patagonia, 1865-1885: a study in historical geography, *Geographical Journal,* **132**, pp. 16-32.

Brading, D. A. 1973. *Los orígenes del nacionalismo mexicano.* Mexico: Sepsetentas.

Brading, D. A. 1980. (ed.) *Caudillo and Peasant in the Mexican Revolution.* Cambridge: Cambridge University Press.

Bryden, J. M. 1973. *Tourism and development.* Cambridge: Cambridge University Press.

Burrus, E. J. 1954. Jesuit exiles, precursors of Mexican independence?, *Mid-America,* **36**, pp. 161-75.

Bushnell, D. 1954. *The Santander regime in Gran Colombia.* Newark: University of Delaware Press.

Butland, G. J. 1957. *The human geography of southern Chile.* London: George Philip.

Buttimer, A. & J. Seamon. 1980. *The Human experience of space and place.* London: Croom Helm.

Caldeira, T. P. 1986. Houses of Respect. Paper presented at Latin American Studies Association meeting, Boston.

Campos H. F. 1960. *Desarrollo educacional, 1810–1960.* Santiago, Chile: Editorial Educo.

Carrasco, P. 1961. The civil-religious hierarchy in mesoamerican communities pre-Spanish background and colonial development, *American Antiquity,* **63**, pp. 483-97.

Carrasco, P. 1972, La casa y hacienda de un señor Halhuica, *Estúdios de Cultura Nahuatl*, **X**, pp. 225-44.

Carvalho, J. M. de. 1980. *A construção da ordem. A elite politica imperial*. Rio de Janeiro: JZE.

Castellanos de Sjostrand, M. E. 1975. La población de Venezuela. Migraciones internas y distribución espacial, 1908-1935. *Semestre Histórico*, **1**, pp. 5-62.

Chávez Ballón, M. 1970, Ciudades Incas: Cusco, capital del imperio, *Wayka*, **3**, pp. 1-14.

Cline, H. F. 1949. Civil congregation of the Indians in New Spain, 1598-1606, *Hispanic American Historical Review*, **29**, pp. 349-69.

Clissold S. & A. Hennessy 1968, Territorial disputes, In *Latin America and the Caribbean*, C. Velíz (ed.). pp. 403-12, London: Anthony Blond.

Cockroft, J. D. 1983. *Mexico: class formation, capital accumulation, and the state*. New York: Monthly Review Press.

Colombia. 1796. *Archivo Nacional de Colombia*. Sección Policía, Vol. X, fols. 537-703. José Peinado y José Antonio Piedrahita, Alférez Real y Teniente Oficial de Medellín, en pleito por preeminencia de asiento en las ceremonias públicas.

Coll, A. & A. Arend. 1985. *The Falkland War, lessons for strategy, diplomacy and international law*. New York: Allen & Unwin.

Collier, D. 1976. *Squatters and oligarchs*. Baltimore: Johns Hopkins University Press.

Conway, D. & J. Brown. 1980. Intra-urban relocation and structure: low income migrants in Latin America and the Caribbean. *Latin American Research Review*, **15**, pp. 95-126.

Dean, W. 1969. *The industrialization of São Paulo, 1880-1945*. Austin: University of Texas Press.

Deler, J-P. 1981. *Genèse de l'espace equatorien. Essai sur le territoire et le formation de l'état national*. Paris: University of Paris.

Delgado Medina, C. 1984. *La crítica del centralismo y la cuestión regional*. Lima: Mosca Azul.

Demelas, D. 1980. *Nationalisme sans nation ? La Bolivie aux XIX-XX siecles*. Paris.

Doughty, P. L. 1974. Behind the back of the city: 'Provincial' life in Lima, Peru, In *Peasants in cities*, W. Mangin (ed.) . Boston: Houghton Mifflin.

Dykerhoff, U. 1984. Mexican toponyms as a source in regional ethnohistory. In *Explorations in Ethnohistory*, H. R. Harris & H. Prem (eds), pp229-52. Albuquerque: University of New Mexico Press.

Ebbing, J. E. 1965. *Aimara: gramática y diccionario*. La Paz: Editorial Don Bosco.

Eidt, R. C. 1971. *Pioneer Settlement in Northeast Argentina*. Madison: University of Wisconsin Press.

Epstein, D. G. 1973. *Brasilia, plan and reality: a study of planned and spontaneous urban development*. Berkeley: University of California Press.

Evenson, N. 1973. *Two Brazilian capitals: architecture and urbanism in Rio de Janeiro and Brasilia*. New Haven: Yale University Press.

Fals Borda, O. 1956-57. Indian Congregations in the New Kingdom of Granada, *The Americas*, **13**, pp. 331-51.

Farriss, N. M. 1978. Nucleation versus dispersal: the dynamics of population movement in colonial Yucatan. *Hispanic American Historical Review*, **58**, pp. 187-216.

Fishburn, E. 1981. *The portrayal of immigration in nineteenth century Argentine fiction, 1845-1902*. Berlin: Colloquium.

Flores-Galindo, A. 1977. *Arequipa y el Sur Andino, siglos XVIII-XX*. Lima: IEP.

Franca, A. 1956. *A marcha do cafe e as frentes pioneras*. São Paulo: Conselho Nacional de Geografia.

Franco, J. 1969. *Spanish American literature*. Cambridge: Cambridge University Press.

Friederici, G. 1960. *Amerikanistischer Wörterbuch und Hilfswörterbuch für den Amerikanisten*. Hamburg: Cram, de Gruyter.

Gade, D. W. 1974. The Latin American central plaza as a functional space. In *Latin America: Search for Geographic Explanations*. R. J. Tata (ed.), pp. 16–234. Boca Raton: CLAG.

Gamba, V. 1986. *The Falklands/Malvinas War: a model of north-south crisis prevention*. Winchester, MA: Allen & Unwin.

Gasparini, G. 1962. *La casa colonial venezolana*. Caracas: Universidad Central de Venezuela.

Gibson, C. 1964. *The Aztecs under Spanish rule*. Stanford: Stanford University Press.

Gillespie, R. 1982. *Soldiers of Perón: Argentina's Montoneros*. Oxford: Oxford University Press.

Góngora, M. 1975. Urban Social Stratification in Colonial Chile, *Hispanic American Historical Review*, **55**, pp. 421-48.

González, L. 1948, El optimismo nacionalista como factor de la independencia de México, *Estudios de Historiografía Americana*, **12**, pp. 143-68.

Gonzalez, M. J. 1985. *Plantation agriculture and social control in northern Peru, 1875-1933*. Austin: University of Texas Press.

Goodwin, P. B. 1977. The central Argentine railway and the economic development of Argentina, 1854-1881, *Hispanic American Historical Review*, **57**, pp. 626-40.

Gossen, G. H. 1974. *Chamulas in the world of the sun: time and space in a Maya oral tradition*. Cambridge: Cambridge University Press.

Gott, R. 1970, *Rural guerrillas in Latin America*. London: Penguin.

Graham, D. H. & S. Buarque de Holanda 1971. *Migration, regional and urban growth and the development of Brazil*. São Paulo: Instituto de Pesquisas Econômicos.

Gravales, G. 1961. *Nacionalismo incipiente en los historiadores coloniales*. Mexico: UNAM.

Guarda, G. 1978. *Historia urbana del reino de Chile*. Santiago: Editorial Andrés Bello.

Hagerman Johnson, A. 1978. Internal migration in Chile to 1921. Unpublished doctoral dissertation, University of California at Davis.

Hall, C. 1976. *El café y el desarrollo histórico-geográfico de Costa Rica*. San José: Editorial de Costa Rica.

Hall, M. M. & H. A. Spalding, 1986. The urban working class and early Latin American labour movements, 1880-1930. In *Cambridge history of Latin America*, Vol. IV, L. Bethell (ed.), pp. 325-66. Cambridge: Cambridge University Press.

Hardoy, J. E. 1955. Evolución de Buenos Aires en el tiempo y en el espacio. *Revista de Arquitectura*, **40**, (375), pp. 25-84

Hardoy, J. E. & M. E. Langdon 1982. El pensamiento regional en Argentina y Chile entre 1850 y 1930. Unpublished paper.

Harris, R. 1983. Centralization and decentralization in Latin America. In *Decentralization and development. Policy implementation in developing countries*, G. S. Cheema & D. A. Rondinelli (eds). Beverly Hills: Sage.

Harrison, R. 1982. Modes of discourse: the *Relación de Antiguedades deste reyno del Piru*, by Joan de Santacruz Pachacuti Yanqui Salcamaygua. In *From oral to written expression: native Andean chronicles of the early colonial period*, R. Adorno (ed.). Syracuse : FACS.

Hawkins, J. 1984. *Inverse images: the meaning of culture, ethnicity and family in postcolonial Guatemala*. Albuquerque: University of New Mexico Press.

Hernando, D. 1973. Casa y familia: spatial biographies of nineteenth century Buenos Airies. Unpublished doctoral dissertation, UCLA.

Hilton, R. (ed) 1969. *The movement toward Latin American unity*. New York: Norton.

Hirobayashi, L. R. 1986. The migrant village association in Latin America: a comparative analysis, *Latin American Research Review*, **21**, pp. 7-30.

Hoberman, L. S. 1986. Conclusion. In *Cities and societies in colonial Latin America*, L. S. Hoberman & S. M. Socolow (eds). Albuquerque: University of New Mexico Press.

Holmer, N. M. 1960. Indian placenames in South America and the Antilles, *Names*, **8**, pp. 133-49.

Holloway, T. 1980. *Immigrants on the land. Coffee and society in São Paulo, 1886-1934*. Chapel Hill: University of North Carolina Press.

Holsten, J. 1986. The modernist city: architecture, politics and society in Brasilia. Ph.D. dissertation, Yale University.

Humphreys, R. A. 1957. The Caudillo Tradition. In *Soldiers and governments*, M. Howard (ed.). pp. 149-65, London: The Athlone Press.

Humphreys, R. A. 1969. *Tradition and revolt in Latin America*. London.

Isabell, B. J. 1978. *To defend ourselves: ecology and ritual in an Andean village*. Austin: University of Texas Press.

Johnson, J. J. 1980. *Latin America in caricature*. Austin: University of Texas Press.

Joseph, G. M. 1982. *Revolution from without: Yucatan, Mexico and the United States, 1880-1924*. Cambridge: Cambridge University Press.

Judge, J. & J. L. Stanfield 1986. The Island of Landfall. *National Geographic Magazine*, **170**, pp. 566-99.

Karnes, T. L. 1983. *The failure of union: Central America, 1824-1960*. Chapel Hill: University of North Carolina Press.

Karrttunen, F. 1983. *An analytical dictionary of Nahuatl*. Austin: University of Texas Press.

Katz, F. 1976. Peasants in the Mexican Revolution of 1910. In *Forging nations: a comparative view of rural ferment and revolt*, J. Spielberg & S. Whiteford (eds). pp. 61–85. East Lansing: Michigan State University Press.

Kemper, R. 1977. *Migration and adaptation: Tzintzuntzan peasants in Mexico City*. Beverly Hills: Sage.

Kicza, J. 1983. *Colonial entrepreneurs: families and business in Bourbon Mexico City*. Albuquerque: University of New Mexico Press.

Klaren, P. F. 1973. *Modernization, dislocation and Aprismo: origins of the Peruvian Aprista party, 1870-1932*. Austin: University of Texas Press.

Kohl, J. & J. Litt 1974. *Urban guerrilla warfare in Latin America*. Boston: The MIT Press.

Korol, J. C. & H. Sabato. 1979. *'The camps': inmigrantes irlandeses en la provincia de Buenos Aires, 1870–1890*. Buenos Aires: Paidos.

Korol, J. C. & H. Sabato 1981. *Como fue la inmigración irlandesa en la Argentina*. Buenos Aires: Paidos.

Kuznesof, E. & R. Oppenheimer (eds) 1985. The Latin American family in the nineteenth century. Special issue of *Family History*, **10**.

Lafaye, J. 1976. *Quetzalcoatl and Guadalupe: the formation of Mexican national consciousness, 1531-1815*. Chicago: University of Chicago Press.

Laite, J. 1981. *Industrial development and migrant labor in Latin America*. Austin: University of Texas Press.

Leeds, A. 1973. Locality power in relation to supralocal power institutions. In *Urban anthropology: cross cultural studies of urbanization*, A. Southall (ed.), pp. 15–41. New York: Oxford University Press.

Leff, N. H. 1982. *Underdevelopment and development in Brazil*. London: Allen & Unwin.

León-Portillo, M. 1984. Mesoamerica before 1519. In *Cambridge history of Latin America*, L. Bethell (ed.), Vol. **1**. pp. 3–36. Cambridge: Cambridge University Press.

Levine, R. M. 1978. *Pernambuco in the Brazilian federation, 1889-1937*. Stanford.: Stanford University Press

Lira, J. A. 1944. *Diccionario Kkechuwa-Español*. Tucumán: Universidad Nacional de Tucumán.

Lobo, S. 1982. *A house of my own: social organization in the squatter settlements of Lima, Peru*. Tucson: University of Arizona Press.

Lockhart, J. 1976. Capital and Province, Spaniard and Indian: the example of late-16th century Toluca. In *Provinces of early Mexico*, I. Altman & J. Lockhart, (eds). Los Angeles: UCLA.

Lockhart, J. 1985. Some Nahua concepts in postconquest guise, *History of European Ideas*, **6**, pp. 465–82.

Lockhart, J. & S. B. Schwartz 1983. *Early Latin America*. Cambridge: Cambridge University Press.

López-Baralt, M. 1979. La persistencia de las estructuras simbólicas andinas en los dibujos de Guamán Poma de Ayala. *Journal of Latin American Lore*, **5**, pp. 83–116.

Love, J. L. 1980. *São Paulo in the Brazilian federation, 1889–1937*. Stanford: Stanford University Press.

Lovell, W. G. 1985. *Conquest and Survival in Colonial Guatemala*. Kingston: Queen's-McGill University Press.

Lowenthal, D. 1985. *The past is a foreign country*. Cambridge: Cambridge University Press.

Lynch, J. 1973. *The Spanish American revolutions, 1808-1826*. New York: W. W. Norton.

Lynch, J. 1981. *Argentine dictator: Juan Manuel de Rosas, 1829-1852*. Oxford: Oxford University Press.

Magallanes, M. V. 1973. *Los partidos políticos en la evolución venezolana*. Caracas: UCV.

Málaga Medina, A. 1975. Las reducciones en el virreinato del Perú (1532-1580). *Revista de Historia de América*, **80**, pp. 9-45.

Mallon, F. E. 1983. *The defense of community in Peru's central highlands: peasant struggle and capitalist transition, 1860-1940*. Princeton: Princeton University Press.

Mayo, J. 1987. *British merchants and Chilean development, 1851-1886*. Boulder, CO: Westview.

Mayer, J. 1973. *La revolution mexicaine*. Paris: A. Colin.

Mead, W. R. 1954. The Language of Place, *Geographical Studies*, **1**, pp. 63-8.

Montoya, R. 1981. *Capitalismo y no capitalismo en el Perú: Un estudio histórico de su articulación en un eje regional*. Lima: Mosca Azul.

Moreno Toscano, A. 1969. Toponimia y análisis histórico, *Historia Mexicana*, **XIX**, pp. 1-10.

Moreno Toscano, A. (ed.) 1978. *Ciudad de México: ensayo de construcción de una historia*. Mexico: INAH.

Mörner, M. 1953. *The political and economic activities of the Jesuits in the La Plata Region*. Stockholm: Almquist & Wiksell.

Morse, R. M. 1958. *From community to metropolis: a biography of São Paulo*. Gainsville: University of Florida Press.

Morse, R. M. 1978. Cities and society in 19th century Latin America: the illustrative case of Brazil. In *Urbanization in the Americas from its beginnings to the present day*, R. Schaedel *et al.* (eds). The Hague: Mouton.

Murra, J. 1975. *Formaciones económicas y políticas del mundo andino*. Lima: IEP.

Murra, J. 1984. Andean societies before 1532. In *Cambridge history of Latin America*, L. Bethell (ed.). **1**, pp. 59–90, Cambridge: CUP, .

Needell, J. D. 1983. Rio de Janeiro at the turn of the century: modernization and the Parisian ideal, *Journal of Interamerican Studies*, **25**, pp. 83–103.

Newton, R. C. 1977. *German Buenos Aires, 1900-1933: social change and cultural crisis*. Austin: University of Texas Press.

Nichols, M. 1968. *The Gaucho*. New York: Gordian Press.

Nuttall, Z. 1921-22. Royal ordinances concerning the laying out of new towns, *Hispanic American Historical Review*, **4**, pp. 743-53; **5**, pp. 249–54.

O'Brian, T. F. 1982. *The nitrate industry and Chile's critical transition, 1870-1891*. New York: New York University Press.

Oliver–Smith, A. 1986. *The martyred city: death and rebirth in the Andes*. Albuquerque: University of New Mexico Press.

Oszlak, O. 1981. The historical formation of the state in Latin America: some theoretical and methodological guidelines for its study. *Latin American Research Review*, **16**, pp. 3–32.

Park, J. W. 1985. *Rafael Nuñez and the politics of Colombian regionalism, 1863-1886*. Baton Rouge: Louisiana State University Press.

Parsons, J. J. 1968. *Antioqueño colonization in Western Colombia*. Berkeley: University of California Press.

Pease, F. (ed.) 1985. *Nueva crónica y Buen Gobierno de Felipe Guamán, Poma de Ayala*. Caracas: Biblioteca Ayacucho.

Pederson, N. 1966. *The mining industry of the Norte Chico, Chile*. Evanston: Northwestern University.

Pike, F. B. 1973. *Spanish America, 1900-1970: tradition and social innovation*. London: Thames & Hudson.

Pitt–Rivers, J. A. 1961. *The people of the Sierra*. Chicago: University of Chicago Press.

Platt, T. 1978. Symétries en miroir. Le concept de *yanantin* chez les Macha de Bolivie. *Annales, ESC*, **33**, pp. 1101–12.

Portes, A. 1979. Housing policy, urban poverty and the state: the *favelas* of Rio de Janeiro, 1972–1976. *Latin American Research Review*, **14**, pp. 3–24.

Ramos, D. 1979. Vila Rica: profile of a colonial Brazilian urban center. *The Americas*, **35**, pp. 495–526.

Ramos Mejía, J. M. 1899. *Las multitudes argentinas*. Buenos Aires: Editorial de Belgrano.

Rausch, J. M. 1984. *A tropical plains frontier: the Llanos of Colombia, 1531–1831*. Albuquerque: University of New Mexico Press.

Raymond, J. 1952. The Indian mind in Mexican toponyms. *América Indígena*, **XII**, pp. 205–16.

Reina, R. 1973. *Paraná: social boundaries in an Argentine city*. Austin: University of Texas Press.

Relph, E. C. 1976. *Place and placelessness*. London: Pion.

Rey Balmaceda, R. 1961. Geografía histórica de la Patagonia. Unpublished doctoral dissertation, University of Buenos Aires.

Ricard, R. 1933. *La conquête spirituelle du Méxique. Essai sur l'apostolat et les methodes missionaires des Ordres Mendiants en Nouvelle Espagne, de 1523 a 1572*. Paris: Presses Universitaires de France.

Roberts, B. R. 1974. The interrelationship of city and provinces in Peru and Guatemala. *Latin American Urban Research*, **4**, pp. 207–35.

Robertson, D. 1978. A behavioural portrait of the Mexican plaza principal. Unpublished doctoral dissertation, Syracuse University.

Robinson, D. J. 1969, Cultural and historical perspective in areas studies: the case of Latin America. In *Trends in Geography*, R. U. Cooke & J. H. Johnson (eds)., pp. 253–267. London: Pergamon.

Robinson, D. J. 1973. Explotación de oro y su impacto en el panorama cultural de la Guayana venezolana en el siglo XIX. *Boletín de la Academia de Ciencias Naturales* (Caracas), **31**, pp. 61–111.

Robinson, D. J. 1975. The syndicate system of the Catalan Capuchins of colonial southeast Venezuela. *Revista de Historia de América*, **79**, pp. 63–76.

Robinson, D. J. 1979a. From colonial space to place. In *Social fabric and spatial structure in colonial Latin America*. pp. 22–24. Ann Arbor: UMI,.

Robinson, D. J. 1979b. Córdoba en 1779: ciudad y campaña, *Gaea*, **17**, pp. 279–312.

Robinson, D. J. (ed.) (in press.). *Population migration in colonial Latin America*. Cambridge: Cambridge University Press.

Rojas Samanez, A. 1985. *Partidos políticos en el Perú*. Lima: Ediciones F & A.

Roseberry, W. 1983. *Coffee and capitalism in the Venezuelan Andes*. Austin: University of Texas Press.

Ruiz, R. E. 1976. *Labor and the ambivalent revolutionaries: Mexico 1911–1923*. Baltimore: Johns Hopkins University Press.

Sack, R. D. 1980. *Conceptions of space in social thought*. London: Macmillan

Sack, R. D. 1986. *Human territoriality*. Cambridge: Cambridge University Press.

Sargent, C. 1974. *The spatial evolution of greater Buenos Aires, 1870–1930*. Tempe: University of Arizona Press.

Schmitt, H. C. 1969. *The roots of lo Mexicano: self and society in Mexican thought, 1900–1934*. College Station, TX: Texas State University.

Schwartzman, S. 1973. Regional cleavages and political patriarchalism in Brazil. Unpublished doctoral dissertation, University of California at Berkeley.

Scobie, J. R. 1964. *Revolution on the pampas: a social history of Argentine wheat*. Austin: University of Texas Press.

Scobie, J. R. 1972. Buenos Aires as a commercial bureaucratic city, 1880–1910. *American Historical Review*, **77**, pp. 1035–73.

Scobie, J. R. 1974. *Buenos Aires: from plaza to suburb, 1870-1910*. New York: Oxford University Press.

Seckinger, R. 1984. *The Brazilian monarchy and the South American republics, 1822–1831: diplomacy and state building*. Baton Rouge: Louisiana State University Press.

Simpson, L. B. 1952. *Many Mexicos*. Berkeley: University of California Press.

Slatta, R. W. 1983. *Gauchos and the vanishing frontier*. Lincoln: University of Nebraska Press.

Solberg, C. 1970. *Immigration and nationalism: Argentina and Chile, 1890–1914*. Austin: University of Texas Press.

Sopher, E. F. 1980. *From pale to pampa: the Jewish immigrant experience in Buenos Aires*. New York: Holmes & Meier.

Spalding, H. A. 1972. Education in Argentina, 1890–1914: the limits of oligarchical reform. *Journal of Interdisciplinary History*, **3**, pp. 41–53.

Spalding, K. 1984. *Huarochirí: an Andean society under Inca and Spanish rule*. Stanford: Stanford University Press.

Specker, J. 1953. *Die missionmethode in Spanisch-Amerika im 16 jahrhundert*. Cologne: Schoneck–Beckenried.

Stabb, M. S. 1967. *In quest of identity: patterns in the Spanish American essay of ideas, 1890–1960*. Chapel Hill: University of North California Press.

Stern, S. 1982. *Peru's Indian peoples and the challenge of Spanish conquest: Huamanga to 1640*. Madison: University of Wisconsin Press.

Szuchman, M. 1977. The limits of the melting pot in Córdoba, 1869–1909, *Hispanic American Historical Review*, **57**, pp. 24–50.

Szuchman, M. 1980. *Mobility and integration in urban Argentina: Córdoba in the liberal era*. Austin: University of Texas Press.

Takagi, H. 1970, The plaza and its function in a Mexican highland community: Tepeojuma. *Geographical Review of Japan*, **43**, pp. 22–31.

Thorndike, G. 1983. *Uchuraccay: testimonio de una masacre*. Lima: Mosca Azul.

Todorov, T. 1982. *The conquest of America: the question of the other*. New York: Harper & Row.

Torre Revello, J. 1945. La casa y el mobilario en Buenos Aires colonial. *Revista de la Universidad de Buenos Aires*. **3**, pp. 285–300.

Tuan, Yi Fu 1974. *Topophilia*. NJ: Prentice Hall.

Tuan, Yi Fu 1975. Place: an experiential perspective. *Geographical Review*, **65**, pp. 151–6.

Tuan, Yi Fu 1977. *Space and place: the perspective of experience*. Minneapolis: University of Minnesota Press.

Tuan, Yi Fu 1984. In place, out of place. *Geoscience and Man*. **24**, pp. 3–10.

Turner, J. F. 1977. *Housing by people*. New York: Pantheon Books.

Uzzell, D. 1974. The interaction of population and locality in the development of squatter settlements in Lima, *Latin American Urban Research*, Vol. **4**, pp. 113–34.

Vayssiere, P. 1980. *Un siècle de capitalisme minier au Chile, 1830–1930*. Paris: A. Colin.

Vanden, H. E. 1979. Mariátegui: marxismo, comunismo, and other bibliographical notes. *Latin American Research Review*, **14**, pp. 61–86.

Vial Correa, G. 1966. La formación de nacionalidades hispano-americanas como causa de la independencia. *Boletín de la Academia Chilena de Historia*, **33**, pp. 110–44.

Wachtel, N. 1971. Pensée sauvage et acculturation. L'éspace et le temps chez Felipe Guaman Poma de Ayala. *Annales. ESC*, **41**, pp. 793–840.

Wasserman, M. 1984. *Capitalists, caciques and revolution: elite and foreign enterprise in Chihuahua, 1854–1911*. Chapel Hill: University of North Carolina Press.

Weibel, L. 1948. Place names as an aid to the reconstruction of the original vegetation of Cuba. *Geographical Review*, **33**, pp. 376–96.

Weinstein, B. 1982. Brazilian regionalism, *Latin American Research Review*, **17**, pp. 262–76.

Weinstein, B. 1983. *The Amazon rubber boom, 1850–1920*. Stanford: Stanford University Press.

White, S. 1981. Movements in the cultural landscape of highland Peru. Unpublished doctoral dissertation, University of Wisconsin, Madison.

Whitehead, L. 1973. National power and local power: the case of Santa Cruz de la Sierra, Bolivia. *Latin American Urban Research*, **3**, pp. 23–48.

Whiteford, S. 1981. *Workers from the north: plantations, Bolivian labor and the city in northwest Argentina*. Austin: University of Texas Press.

Williams, R. Bryn 1964. *Y Wladfa*. Cardiff: Gwasg Prifysgol Cymru.

Williams, R. B. 1976. The structure and process of Welsh emigration to Patagonia. *Welsh History Review*, **8**, pp. 42–74.

Wirth, J. D. 1977. *Minas gerais in the Brazilian federation, 1889–1937*. Stanford: Stanford University Press.

Woll, A. 1982. *A functional past. The uses of history in nineteenth century Chile*. Baton Rouge: Louisiana State University Press.

Womack, J. 1969. *Zapata and the Mexican revolution*. New York: Knopf.

Wortman, M. 1976. Legitimidad política y regionalismo. El imperio Mexicano y Centroamérica. *Historia Mexicana*, **26**, pp. 238–62.

Zuidema, R. T. 1964. *The Ceque system of Cuzco: the social organization of the capital of the Incas*. Leiden: E. J. Brill.

11

The power of place in Kandy, Sri Lanka: 1780–1980

JAMES S. DUNCAN

Introduction

It is always a matter of some considerable regret when there exists a gulf between subdisciplines which potentially have much to offer one another. This is particularly true when the phenomena which each studies are intimately related in real life, and it is merely some quirk in the development of an academic discipline which separates them. Such has been the case with the separation of cultural geographers interested in place and the interpretation of landscapes, from political geographers with their interest in political process. In a sense these two subdisciplines within geography have stood on different sides of the great divide in post war geography, with the cultural geographers focusing upon the environment, stressing the artifactual and looking toward either the humanities or a social science which is idiographic in orientation, and the political geographers focusing on space and looking to a more nomothetic view of social science. The problem for geography, it would appear, is that cultural geographers have had too narrow a view of culture while political geographers have had too narrow a conception of politics. Fortunately it has become increasingly apparent that a radical separation of politics from culture and place diminishes all three concepts, for it creates, on the one hand, an impoverished view of political power as abstracted from the cultural and locational matrix of which it is so clearly a part, and on the other hand a view of culture and place which ignores the importance of political processes in both cultural continuity and change in place. The importance of the politics of culture has long been espoused by Clifford Geertz (1973, 1980) and has been reinforced of late by a growing interest in North America in the work of such European thinkers as Antonio Gramsci (1971) and Raymond Williams (1973). In particular, their concepts of ideology and hegemony have helped forge links between culture and politics. It remains for geographers to incorporate the concepts of landscape and place.

This chapter will focus upon the interrelationship between four concepts: ideology, political power, landscape, and place. All will be examined within the context of a cultural paradigm. Ideology is a set of beliefs about how social life is or should be organized. Examples of ideologies are equality of opportunity, caste, the divine right of kings, or the right of national self-determination. The term ideology is used here in a non–evaluative sense;

that is, no a priori judgment is made about the truthfulness or goodness of the beliefs. Politics includes not only overt decision making on the part of governments, but also includes the ideological frameworks which shape political life and legitimize institutions of governance. It includes styles of maneuvering and mobilizing power within particular political structures as well as the modes of restructuring power relations. This manipulation of ideological frameworks is both constrained and enabled by the larger cultural paradigm of which it is a part.

A landscape can be distinguished from the more general term environment which is composed of the objects that we encounter in the world: hills and valleys, trees and fields, towns and villages, houses and streets. A landscape, however, is a culturally produced model of how the environment should look. It is, therefore, not merely an environment but a type of arrangement of hills and trees, or towns and houses. Environments become transformed into landscapes as people transform them physically or merely reinterpret them in such a way as to bring the environment in line with a particular landscape model. These models have cultural and historical specificity so that one can, for example, speak of the landscape model of a royal capital in 18th century Indian Asia, the rural romantic landscape of 19th century England, or the suburban landscape of mid-20th century America and by these terms indicate generally what the arrangement of objects in the environment in each case would be. These landscape models are complex because they can escape their original cultural and historical origins. One can find, for example, 19th century rural romantic English landscape models being superimposed upon 18th century Indian ones, or 19th century English ones combined with late 20th century American suburban styles. When they are wrenched from their cultural and historical contexts landscape models may have heightened ideological and often political significance.

Place is defined here as a specific locality, which is characterized in a significant way by its landscapes. In this chapter I intend to focus upon the city of Kandy and show that it is characterized by a number of different landscapes which are created and modified over time as a part of the cultural and political changes taking place there. Just as places have cultural and political significance, so do landscapes. I will argue that much of the power of a place derives from the power of the landscape model of which it partakes.

The city of Kandy is nestled in a valley surrounded by low mountains in the central mountainous core of Sri Lanka. It has been important within Sri Lanka since the 15th century not only because it was the capital of the Kandyan kingdom but because in the early 16th century the most important relic in Buddhism, a tooth of the Buddha snatched from the ashes of his funeral pyre in the 6th century B.C. was brought to Kandy by the king and enshrined in a royal temple adjoining the palace. As such, Kandy was and remains an important place of pilgrimage. But Kandy also derives its importance from the fact that for over 300 years it successfully repelled European invaders, first Portuguese, and then Dutch armies until it finally succumbed to a combination of the might and guile of imperial Britain in the early 19th century. From the fall of the last of the coastal kingdoms of Sri Lanka until the present time, Kandy has served for Sri Lankans as a symbol of nationalism and of resistance to western imperialism.

In this chapter the political significance of Kandy under three different cultural paradigms will be discussed: first, the Kandy of the god-kings in the late 18th century, second the Kandy of the British in the 19th century, and third the Kandy of the present government in the early 1980's. The meaning of the place has changed during each of these periods and yet a central narrative structure runs through Kandy linking past to present.

Pre-colonial Kandy

Late 18th century Kandy was an archetypal South Asian capital. It was the city of the god-king, the center of the universe. It was Mount Meru, the city of the gods descended to earth, and the stage upon which the king could display power. The city of Kandy became the capital of an independent state in the mountainous core of the island in the late 15th century. The Tooth Relic of the Buddha, the palladium of kings, was brought to the city in the 16th century, lending much prestige to the capital and the king who occupied the throne.

The city was composed of two rectangles, the rectangle being the sacred shape of the cosmic cities of the gods. The western rectangle constituted the main city and contained both the residences of the nobles and the houses and shops of the common people. This rectangle was in turn divided into 21 squares representing the 21 administrative units of the kingdom. The kingdom, therefore, was symbolically reproduced within the city; the macrocosmos was reduced to the microcosmos. When the king annually circumambulated the city in ritual procession he reaffirmed his control not only over the city but over his whole kingdom as well. This ritual act should not be seen as being *merely symbolic* in the sense of being non–efficacious. Circumambulation with the right side facing the city was a ritual act which was deemed to have *magical* efficacy in maintaining power over the capital and the wider kingdom that was magically reproduced within it through the power of like numbers.

But to really understand the role that the landscape of this place played in the legitimation of power we must understand the meaning of the eastern rectangle, the so called "sacred square". This eastern rectangle contained the temples and the palace and was the real locus of ritual power in the kingdom.

To the south of the sacred square lay a lake called the Ocean of Milk. This is the name in the sacred texts given to the cosmic ocean which lies at the foot of Mount Meru at the center of the universe.

The sacred square contained the palace of the king, the Temple of the Tooth Relic of the Buddha, and three temples to the gods. Within the Indian tradition of royal cities, the sacred square normally represents Mount Meru, the sacred cosmic mountain at the center of the world. On the eastern side of the sacred square was located the Temple of the Relic and the palace of the king joined together in a walled compound. This palace/ temple complex is identified in 18th century Kandyan court poetry as Mt Mandara which is the eastern peak of Mt Meru (Godakumbura 1961). This then places the capital at the center of the world and further accounts for the location of the palace/temple complex in the eastern portion of the center. The king of Kandy is continually referred to in the *Culavamsa* (1953) (the official chronicle of the kingdom) as being like Saccra the king of the gods and it is interesting to note that in the Puranic texts the king of the gods' city is said to be located on Mt Mandara, the eastern peak of Mt Meru (Dimmitt & Van Buitenen 1978). The identification of the palace with the Temple of the Relic is very close. These two structures mirror each other in terms of their spatial layout and the names of functionaries (Seneviratne 1978). The king showed spatially and through the duplication of functionaries his close identification with the relic of the Buddha which gave his rule a supernatural power. Around the Temple of the Relic was a wave wall which symbolized the heavenly Ganges descending from Mt Mandara and flowing into the Ocean of Milk, and above this is the cloud wall or Celestial Rampart which symbolizes the heaven of Saccra, the king of the gods.

At the northern edge of the sacred square was the temple of the god Visnu. Post-Vedic mythology places the heaven of Visnu in Vaikuntha the northern peak of Mt Meru. At the southern edge of the sacred square was the temple of Nata (the Maitreya or future Buddha) whose location according to the Buddhist texts is on the top of Mt Meru. At the western edge of the sacred square is the temple of Pattini a manifestation of Parvati the consort of Shiva who also dwells on the top of Meru. The temples of the four gods, therefore, filled out the cardinal directions of the sacred square and made it a complete replica of heaven on the top of the cosmic mountain. At the center of the sacred square there was a Bo tree grown from a shoot taken from the tree under which the Buddha was enlightened. The Bo tree, for Buddhists, is a cosmic axis and partakes of the symbolism of the central mountain. In the center was also found a stupa which contained a relic of the Buddha. Stupas are symbolic of Mt. Meru and are therefore thought to have a cosmic axis running through them.

The cardinal directions also imbue the sacred and profane squares with a moral quality. The east for example is "higher" in terms of ritual purity than is the west. It follows that the further we move east in Kandy, the more elevated we are morally. For example, to the east of the profane square is the sacred square. To the east of the sacred square is the Temple of the Relic. In the eastern part of the Temple of the Relic is the Shrine of the Relic, and in the eastern part of the Shrine is the Relic itself. We can see then that urban form in Kandy speaks of hierarchy, of power, and of the relationship between mere mortals and the god-king. The city as a whole represents the cosmos in miniature with the profane square containing the citizens symbolizing the earth and the sacred square containing the Buddha, the gods and the king symbolizing heaven.

What we have seen here in the Kandyan landscape is a political and religious argument made in the vocabulary of ritual; that political power is but a transformation of divine power. The layout of the town concretizes the ideology of the god king. The king, by inserting himself into the center of this mythologically charged place, makes immanent the ideology. He acts out his role as a charismatic ruler inextricably linked to the animating center of the society. The charismatic god king, precisely because he is enshrined in this heaven on earth, causes the state to appear to have a liminal quality – suspended between the heavens and the people.

How can one gauge the political impact of the landscape of Kandy? It is indeed difficult, but not impossible, to ascertain the perceptions of people living 200 years ago. There are, in the case of Kandy, written texts that can be used to answer this question. For example there were a number of Indian texts that were well known in Kandy which held up the ideal of the god-king in his heavenly city. There were the *Puranas*, the *Nikayas*, the Arthastra, and the Silpa literature. There were also approving references to Kandy as a city of the gods in the *Culavamsa* (1953), the Great Chronical of Ceylon, and in late 18th century Kandyan court literature (Godakumbura 1961). We also have folk tales and rituals from this period (Dolapihilla 1959, Obeyesekere 1984) and the reports from British spies which reaffirm this view (D'Oyly 1917). However, there is also strong evidence that the last king's concern for the landscape of Kandy contributed to his downfall, for he engaged in such massive building programs in Kandy between 1811 and 1814 that he alienated his people. This suggests that while symbolism is a powerful tool for reinforcing power it does not guarantee power, for the last king of Kandy was overthrown at the beginning of the 19th century, betrayed to the British by his own nobles. He tried to curb the power of the nobles and stop their treasonous intrigues with the British by instigating a reign of terror against some of the leading families of the kingdom, but ultimately the power of the united noble factions and the British military power was too great and he was defeated. The nobles turned to the British to help them remove one particular king whose rule they found onerous. They hoped that the British would then leave and allow them to appoint a more pliable monarch. The British, however, were intent on achieving a political transformation and reducing the Kandyan State to a part of the British Empire. The political transformation produced a new cultural paradigm which was distinct from that of the Kandyan kingdom and in the process transformed the landscape of Kandy.

Colonial Kandy

Although the British in 1815 had military control of the capital and the nominal support of the nobility and the population they were faced with the problem of establishing political legitimacy in a society whose cultural/political paradigm was profoundly different from their own. Something had to be done to achieve legitimacy, for if the gap between political ideology and political practice becomes too great, the latter is rendered problematic. The problem for the British was that the Kandyan political ideology was one of the incorporation of the ruler into a charismatic brand of state-sponsored Buddhism. Such a position was structurally impossible for the British on two counts. First, the role of the charismatic king actively engaged in court rituals was impossible for an English king who lived on the other side of the world, and second the English king was the head of the Church of England which was interested in proselytizing not assimilating to Buddhism. And yet knowing this full well, the British upon their entrance into Kandy in 1815 signed a treaty promising to maintain the traditional relationship between the ruler and the religion. The governor and government agent even went so far as to honor the Buddhist monks and make the traditional offerings to the Tooth Relic. John D'Oyly, the government agent (D'Oyly 1917) cynically referred to this in his diary as "a little political humbug." By 1818 the Kandyan aristocracy realized that the whole political paradigm was under attack and revolted. The British used this revolt as a pretext both to revoke the treaty of 1815 and crush the power of the nobles as they had crushed the king three years earlier. Although the Tooth Relic had the power to mobilize the masses, the British were unable to use it, for its power stemmed from the oneness of the ruler and Buddhism. There was intense pressure on the colonial government from Christian missionary groups to disassociate itself from the relic. Over the objections of practicers of *realpolitik* such as the government agent D'Oyly, the government progressively distanced itself from the relic. But in so doing created an ideological vacuum. The basis of charismatic rule in the center of Kandy had been undercut. The British therefore attempted to fill the political vacuum by co-opting some Kandyan symbols and by substituting some of their own.

The British usurped and transformed the Kandyan symbols of authority in a number of ways. There existed a belief in Kandy that no foreign power could rule the kingdom unless three conditions were first met. The foreigners had to take possession of the relic that was jealously guarded by the monks of the Tooth Temple and hidden away in the mountains when foreign armies invaded. They also had to build a road from the coast up through the mountains to Kandy, and finally they had to ride a horse through the heart of a mountain. The British saw these as an impediment to ruling Kandy and immediately set about fulfilling them. Upon their arrival in 1815 they negotiated for the return of the Tooth Relic to the temple where their soldiers could keep an eye on it. They also started work on a road connecting Kandy to the coast. And finally, to the utter amazement of the Kandyans, the British used black African troops to tunnel through a mountain on the outskirts of Kandy so that a coach and four could pass through. A number of years later a British civil servant in Kandy criticized both the existence of the tunnel which he rightly claimed was unnecessary, since that particular mountain could have been easily skirted, and the design of the tunnel itself. He was informed by another civil servant who was in Kandy at the time of the tunnel's constructions that the tunnel "was never intended as a specimen of engineering science but to impress the Kandyans with the full sense of the irresistible power of their conquerors." (Stewart 1862, p. 7)

Upon their arrival in 1815 the British immediately set about replacing the Kandyan power structure and symbols of authority with their own.

They replaced the king of Kandy's troops with their own, placing their native Malay and African troops on the eastern outskirts of the city and their European troops on the western side where they could better keep an eye on the Kandyan populace. The palace of the king became the residence of the government agent, the highest ranking Englishman resident in Kandy. The king's audience hall located on hallowed ground between the Temple of the Relic and the palace became the civic court during the week and the Anglican church on Sundays. In the alcove where the king of Kandy's throne once sat stood the pulpit and behind it hung a picture of the English king. The Malabar palace of the king's relatives became the European hospital and the queen's bath became a European library. The octagon of the sacred Temple of the Tooth Relic became in succession a military jail, a survey office and a library. The large houses of the nobles were taken over after the rebellion of 1818 and converted into stores and hotels. The streets were all renamed, either after natural features such as Hill Street, and Lake Road or after governors and their wives such as Ward Street and Lady Torrington's Drive. The British also created new symbols of authority such as the Governor's Pavilion built in the 1820s for the governor of the island when he visited Kandy. This building dwarfed the palace of the king. The pavilion was, in the words of one contemporary, "calculated for the abode of one who is entrusted with the government of Asiatics" (Sirr 1850, p. 92). Before the completion of the Pavillion a European traveler to Kandy claimed that the two most impressive buildings in the city were the new jail built adjacent to the temple square and the military commandant's house in the western section of the city. It is clear that in the early days of the British occupation the military presence was paramount. The sacred square in the center of Kandy which had been reserved for the Temples of the gods, the Relic, and the palace of the god-king was invaded by British institutions. An Anglican church was built there which towered over the Temple of the Tooth Relic to the dismay of the Buddhist monks who considered it a sacrilege to have any building higher than the Relic. A Protestant school was also placed within the sacred square as were the police courts. The invasion of the sacred square reached its peak when tennis courts were built within the grounds of the Temple of the Tooth in order that the government agent might enjoy a game of tennis in the evening. Only slightly less offensive were the regular cricket and rugger games played on the esplanade in front of the temple. British statues were prominently displayed in public squares. A larger than life statue of Governor Ward was placed at the far end of the esplanade and on the base of the statue was a famous quotation of the good governor referring to the Sinhalese as a "semi-civilized people" (Bingham 1921, p. 122). The other major statue was to the British soldier. There was also a great social separation between the British and the Sinhalese. Clubs and even hotels remained segregated in some instances until World War II.

The British policy of creating new symbols of authority in Kandy went hand in hand with a policy of undermining the old symbols of authority. The parts of the palace that were not occupied by the government agent were allowed to fall into disrepair. In fact mid-19th century visitors often were unaware that it used to be a palace at all and simply referred to it as the agent's bungalow. The temples in the sacred square and elsewhere were denied government sponsored funding and began to disintegrate. The kings' tombs in the northern part of the city were vandalized and finally destroyed when they were chosen as a site for a railway line into Kandy. The destruction of the traditional city in the first decade of British rule was very great. As one British observer in 1820 noted:

> We have pulled down much and built up little; and, in taking no interest in the Temples, we have entirely neglected their repair; the consequence is, that Kandy has declined very much in appearance during the short time it has been in our possession, and to the natives must seem merely the wreck of what it once was. In a few years, in all probability, not a vestige of the old town will remain. . . . (Davy 1821, p. 371)

The British had consciously set out to modify the landscape of Kandy to reflect their values. The symbols of power of the Kandyan king, and the nobles had been eliminated, or taken over and associated with the new rulers. Other new symbols were put in place to reflect the power of the British. What the British couldn't use they let decay, mirroring in the landscape of the place the decay they wished to see in the institutions themselves.

The meaning of Kandy as a place was being changed. A new cultural and political system had been ushered in and the old landscape model which spoke of what had been important under the old system was being transformed or allowed to fall into ruination. Kandy was no longer the city of the god king, for he had been unceremoniously sent into exile. It was becoming a British colonial town and it had to look the part. The British realized that if they were to achieve legitimacy it would have to be largely on their terms. And whilse they never achieved the degree of legitimacy that they sought, they achieved a degree of cultural hegemony among the Sinhalese elites. This cultural hegemony was achieved in part through a conscious attempt to change the Sinhalese elites, but very largely it was achieved through an attempt to transform Kandy into an outlier of British culture. The assimilation was left to the Sinhalese themselves who were often all too ready to emulate the British. This creation of a bit of British culture in Kandy involved a transformation of the landscape of the place and the natural environment of the Kandyan area lent itself admirably to this task. Because of the elevation and topography of Kandy it was possible to create a facsimile of the landscape of home. One could see in residents' diaries, and official plans the conscious attempt to transform Kandy into a hybrid that was part English and part Sinhalese. Their success in doing so was attested to in the journals and paintings of travelers who visited Kandy throughout the 19th century. Kandy was designed to resemble a romanticized image of a pre-industrial England. The landscape model of the English lake district was superimposed upon the mountains and the Kandy lake to recreate a place where English ladies and gentlemen could somehow escape the tropics and the native culture and symbolically return home. And how exactly was this done? Promenades, such as Lady Horton's Walk, carriage drives such as Victoria Drive, and riding paths such as The Green Gallop, were created (see Fig. 11). The dense jungle around Kandy was pruned to reveal the best views of the town, the lake and the surrounding mountains. Travelers and residents alike often wrote about how one might "find enchanting views suddenly opening from the various points where the thick verdure of the trees has been judiciously cut away" (Dougherty 1890, p. 102) and that these openings created an "exquisite framework through which... [to] see the distant landscape" (Cave 1912, p. 303). The term sublime was commonly used to describe the Kandyan landscape. English vegetables, fruits and trees were introduced and both formal and informal English gardens and parks were laid out. The Governor's Pavilion was laid out like an English country house situated in a parklike expanse of lawns and shrubbery overlooking the town, the lake and the mountains. Exclusive European residential quarters were located around the lake. English architect-designed bungalows with their gardens full of roses and other English flowers climbed the hills from the lakeside. English style buildings predominated on the main thoroughfares such as Ward Street which contained the European stores like Cargills, Walker and Company, the Merchantile Bank, the Queen's Hotel, the Kandy Club, the Lawn Tennis Club, and the Planters Association of Ceylon. The upper-middle class sporting life of Britain was transferred to Kandy and dominated the social scene. Kandy fielded sides in cricket and rugger against Colombo

THE CITY OF KANDY

T - Temple of Tooth
A - Audience Hall
N - Nata Temple
P - Pattini Temple
V - Visnu Temple
K - Kataragama Temple
D - Dagoba
B - Bo Tree
KP - King's Palace

MP - Malabar Palace
QB - Queen's Bath
GP - Govenor's Pavillion
J - Jail
MC - Military Commander
AC - Angelican Church
PS - Protestant School
PC - Police Courts
TC - Tennis Courts
KT - King's Tombs

Sacred Square

—— Roads

BOGAMBARA LAKE

OCEAN OF MILK
(KANDY LAKE)

Figure 11.1 The generalized structure of Kandy, Sri Lanka (1780–1980).

193

and other towns, and boasted such island-wide stars as S.K. "Bowser" Bousfield, "Borneo" Jamieson, and "Poochie" Papillon. There was also tennis, golf, racing, polo, regattas on Kandy Lake under the direction of J.D. Aiken and Miss Dicks, hunting deer and jackal with packs of hounds, and then of course there was the premier club of all, the ABCD (Athletics, Boating, Cricket, and Dancing) Club (Wright 1951, p. 72, 75, 231). English schools such as Trinity College were set up to produce Christian natives who would mirror English taste and who could serve as underlings in the bureaucracy. On the outskirts of Kandy were the coffee and tea estates. Whilst some had Sinhalese names the majority had names like Eton, Pine Hill, Gloucester, Wiltshire, Windsor, Prospect Hill, and Roseneath (Ferguson, N.D.). The diary of a planter from 1838 is interesting in this regard (Boyd 1889, p. 17):

"What's the name of the place?" asked Braybrooke the Government Surveyor. "Oh! Ah! Yes! Dodang-gaha-Kellie," he replied; "Very long name but not very pretty, although tolerably expressive I daresay. Now, I for one, don't admire those native names. People say they are expressive, but expressive of what I should like to know, if you don't happen to understand the language? When I come into my landed property, I intend to call it Oxgrove or Hogg Lodge, or by some other good old Saxon name. I hate these native languages. A gentleman has no use to learn more of them than enough to swear fluently at a native with."

Kandy in the late 19th century had to some become so like a bit of England that it was disquieting (Hurst 1890, p. 202).

In Kandy, whether one will or not, the mind will go back to the Lake region in England. You find a calm and quiet beauty, freedom from strain and stress, a cluster of hillsides which throw down their beautiful face into the mirroring lake at their feet, a sweetness in all the pulsations of the air and a universal friendliness between all Nature and its lord, which brings up Grassmere, Windermere, Derwentwater, and their spirits – Southey, Wordsworth, Coleridge, and all the rest of the Cumberland immortals. Even the hostelry of Kandy, the Queen's Hotel, suggested to me immediately the Keswick Inn. The whole reminder was pleasant enough at first. But too much of such resemblances is not good. You can less easily resign yourself to the novelty of your new environment. You lose the new by recalling too intensely the old.

Nearly a century later in the 1960s the resemblance to an English landscape was still striking. The late theologian Thomas Merton stayed at the Queen's Hotel in Kandy and wrote (Merton 1968, p. 216), "Now I find myself looking out the hotel window at an inexplicable English village church up against what might, but for a couple of coconut palms, be a Surrey hillside."

Clearly Kandy was no longer the same place in 1840 that it had been in 1815 when the English captured the city. The landscape model had been radically altered. The political symbolism of the god-king formerly so deeply encoded in the landscape was replaced by the symbolism of the English country gentleman with his garden, his clubs and his sport. This new model had a powerful hegemonic effect upon members of the Kandyan elite. Many ambitious Kandyans converted to Christianity, and culturally transformed themselves into "brown Englishmen". By the late 1830s some Kandyan chiefs began to follow "European models" (Forbes 1840, p. 123). One commentator (Pridham 1849, p. 263) at this time noted that "Many of the Chiefs have latterly had houses built after the European manner, and have introduced European luxuries into their internal arrangements." In 1983 I interviewed a member of the Kandyan upper class who was an ardent nationalist. He had this to say about his father during the waning years of the British period.

He aped the British in virtually everything. I remember seeing him sitting alone in full dinner dress at dinner being served by the butler. In the pre-war years if the government agent came to tea at your bungalow it gave you a certain standing, and if the Governor stopped at the house on his way to Nuwara Eliya, then that really made you someone. You became known both among the British but more especially among the Sinhalese as someone to be contended with. Look at that picture of my mother as a little girl on the wall. From a distance, if you didn't see she was brown skinned, you would think she was a little English girl in a summer dress. That is what we were, "Brown Englishmen." Our culture was nearly destroyed by this attitude and yet many of us in the elite willingly embraced it and sought it out. We accepted the English vision of themselves as superior and their standoffishness helped promote this since in reality many were uneducated louts – especially the Planters. But people blindly looked up to them. My father saw his knighthood as the culmination of his career and a true sign that he had reached the social apex. But to me it was a moment of shame for it represented a Sinhalese capitulation to the British.

This "landscape of home" must not be seen as value neutral in a political sense simply because the English themselves may not have seen it as an explicit political tool. The transformed landscape of Kandy was intensely political because it symbolized English power – the power to command large tracts of land, to ride in the sacred forest, to play tennis within the grounds of the most sacred shrine in the Island. It also symbolized the claimed superiority of English culture, for the landscape tastes, house styles, consumer goods and manners of the elite were foreign to Sinhalese culture.

And yet there are two principal reasons why the hegemonic effect of the English transformation of this place was far from complete. In the first place, only the Sinhalese elites were influenced by the English symbolism and they came to guard this symbolism jealously against the lower classes.

195

For example as late as the mid 1950s English speaking Sinhalese thought it improper for non-English speakers to wear trousers rather than a sarong. Second, romanticism itself, which played such a crucial role in shaping the landscape of the place ultimately undermined the cultural hegemony that it helped create. For the old and the exotic landscape elements which increasingly played a part in the ideology of romanticism, in Kandy were drawn from a pre-colonial cultural tradition. The result was the incorporation of not merely landscape elements but at the same time religious and nationalist feelings symbolized by those elements which had been submerged under a veneer of English culture. It is noteworthy that this "rediscovery" of the romantic and picturesque quality of Kandyan culture took place after the mid-19th century when the Kandyans no longer posed any military threat to English rule. This rediscovery took the form of trying to restore buildings in Kandy that had been neglected by the British earlier in the century. Governors Gordon and Gregory played an active role in the restoration of Kandy in the late 1800s and Governor Blake in the 1900s sponsored a movement to revive and preserve Sinhalese painting and handicrafts. Even new buildings were to have Kandyan detail as ornamentation. Those that didn't came under heavy criticism. For example, in the 1900s the new government agent's office was criticized as "an extensive and handsome building but alas! having no feature of any kind that harmonizes with its surroundings. In an English manufacturing town it would not be out of place; but in Kandy it is a deplorable incongruity" (Cave 1912, p. 312). Even the governor's residence in Kandy came under fire. "It is fine in spaciousness but its furniture is uninteresting English stuff of the kind one finds in a Liverpool hotel, instead of the rich oriental things they might have so easily had... It made me sick to see the cheap Nottingham curtains in the windows where Indian silks and embroideries naturally belong." (Rogers 1903, p. 252) This interest in making Kandy more "oriental" fed into a growing tourist trade, for as early as the 1840s, tourists with "oriental antiquarian" interests (Bennett 1843, p. 395) were coming to Kandy. One can only surmise that there was a conflict of interest between the desire to make Kandy as much like England as possible for the benefit of the local English and as oriental as possible for the tourists. Perhaps an eclectic, picturesque romanticism was the compromise. While romanticism relegitimated Kandyan culture, it trivialized it, by converting it into a museum piece, a culture with a past but no future. Kandyan culture was seen as something having mere decorative value, which could render the landscape more picturesque. The journals of travelers again and again allude to Kandy as a relic of the past. As one traveler put it in 1908 (Farrer 1908, p. 49), "Saddest...of all dead places in dead Lanka is the last capital of the...Cinhalese kingdom."

And yet, even this trivialized form of Kandyan culture kept alive the seeds of nationalist sentiment, for the British interest in Sinhalese culture history helped legitimate it for a Sinhalese elite that had distinct feelings of inferiority. The restoration of cultural monuments helped foster a growing national pride and resistance to English Imperialism which culminated in independence in 1948.

Post-colonial Kandy

Although Sri Lanka achieved its independence from Britain in 1948, the early post-independence governments were secular and elitist in orientation and it was not until the election of 1956 that a prime minister was elected on a populist Buddhist platform (see Bechert 1978). Although the politics of Buddhist revival has continued to have broad appeal to the majority Sinhalese to this day (see Obeyesekere 1972, Phadnis 1976, Seneviratne 1984) it has completely alienated the minority Tamils and to a lesser degree the Muslims (see Tambiah 1986). Within this post-colonial political paradigm Kandy has been transformed from a sleepy colonial hill city to a symbol of a revitalized Sinhalese Buddhist identity. Kandy has reachieved its position of eminence primarily because the relic of the Buddha is enshrined there. Once again the relic has become an important link between the religion and the government. Kandy also is a symbol of Sinhalese pride in that it resisted European domination for 300 years until it finally succumbed, not as much to the might of a British army, as to the perfidy of its own aristocracy. As such, it is seen as a place of Sinhalese cultural purity, which was less polluted by European values than the coastal portions of the island.

Although the symbolic meaning of Kandy has been transformed, much of the landscape of British Kandy has remained intact. Post-1956 governments have had to create their own symbolic landscape within this inherited framework. The government has symbolized the new relationship between place and politics by making certain key changes in the landscape. The most obvious symbols of British domination such as the statues of Governor Ward and the British Soldier were removed and in their place were erected five statues, one of which commemorates the brave opposition of a young boy to the last Tamil king of Kandy in 1814, two of which commemorate Buddhist opposition to the British and two of which commemorate post-independence politicians. Interestingly none of the statues commemorate either the kings of Kandy or the pre-colonial Sinhalese nobility. The problem is that both of these groups are flawed in the eyes of the contemporary Sinhalese politicians, for although the Kandyan kings in the 18th century were nominally Buddhist, they were originally Tamils from the south of India. The Sinhalese nobility, on the other hand, welcomed the British army into Kandy in order to overthrow an unpopular king. The street names also were all converted from British to Sinhalese names. But again, the streets were not given their pre-colonial names. Some of the English names were simply translated into Sinhalese, others were changed to refer to the religion, while others were renamed after post-independence politicians. In 1983 I conducted a survey of 143 Kandyans eliciting their perceptions of the city. Their responses to these changes were enlightening. Whereas 89% adopted the new name of the street named after the relic, only 9% used the new names of any of the three streets named after politicians. Many people were critical of the latter as mere self-aggrandizement or party politics. Official buildings were modified. The king's palace which had become the government

197

agent's bungalow was stripped both inside and out of its English trappings and converted into a Kandyan museum. The Governor's Pavilion, on the other hand, was converted into the President's Pavilion with relatively little change. New official buildings such as the Buddhist Association Building and the National University were built in a neo-Kandyan style. Finally, the sacred square that had been desecrated by the British by placing a Christian church, a school, and even tennis courts in it was rededicated as a sacred square and talks were begun on removing the offending buildings.

This whole process of the symbolic decolonization of the landscape of Kandy has itself been thoroughly politicized. Since independence Kandy has been a stronghold of the conservative United National Party (UNP) and hence that party has had better access to the religious and nationalist symbolism of the place. For example, the statues in town are not merely to nationalist leaders but to important UNP figures, and the streets are named after UNP politicians. During the 1970s when the opposition Sri Lanka Freedom Party (SLFP) was in power they named one street after their party leader Mrs Bandaranaika, but the new street name was seldom used, except as I discovered by one SLFP postman who regularly threw away mail that didn't refer to the street by his beloved leader's name. The major temples to the gods and the relic of the Buddha in Kandy are also highly politicized for they have elected lay leaders who control the administration of the rich temple lands. These positions which carry both honorific and pecuniary rewards are regularly contested by members of both major parties. These temples display both their affluence and influence during the Perahara, that great religious and political festival held during the full moon week in August. A number of years ago there was a great uproar because the one temple controlled by the SLFP managed to secretly hire most of the elephants normally used in the parade so that their temple's procession was very grand and the opposition's ones were rather pathetic by comparison. Similarly after the UNP victory in 1977, the leader of the SLFP Mrs. Bandaranaika was disbarred from politics and in response a great throng of her supporters marched to the temple of the goddess Pattini to seek the goddess' redress in this matter. Needless to say this temple was at this time in the hands of the SLFP.

Perhaps the post-independence politician who is most skilled at manipulating Buddhist imagery is the current president of Sri Lanka, J. R. Jayawardena, or J.R. as he is known to his countrymen. The president not only uses Buddhist imagery, but the imagery of the pre-colonial Buddhist kings of Kandy. After his election the president and his cabinet immediately came up to Kandy where upon entering the city he passed under a triumphal arch with a crown on the top. He then went with his cabinet to the Temple of the Relic to pay homage to the relic as did the kings of old in Kandy. His first address to the nation was made from the octagon of the Temple of the Relic, because he said "It is from this spot that our ancient kings addressed the people" (Jayawardena 1974, p. 130). The president also commissioned a personal flag for himself that is raised on all important occasions when he speaks. This flag is decorated with the "Dhamma Chakra" the wheel of the Buddha which symbolizes the teachings of the Buddha. But the wheel has another meaning as well. It also symbolises the "Chakra vartin" the wheel of the all powerful south Asian king, and it is precisely this symbolism of the pre-colonial buddhist monarch that the president is using. The president has also, like the Sinhalese kings devoted a lot of attention and money, much of which is provided by UNESCO, to the restoration of ruins in Kandy and elsewhere.

However, while looking back to a pre-colonial past, the present government has, since 1977, opened the country up to an unprecedented flood of Western goods and tourists. It appears that the more the country is opened up to the West, the more the rhetoric and symbolism of tradition is invoked. And yet in spite of their symbol mongering the politicians can't return to the pre-colonial past. For Sri Lanka is a late 20th century Asian country which still feels the deep imprint of its colonial past. One can see the tension between the present, the colonial and the pre-colonial past in the landscape of Kandy. One can see it in schoolboys in white playing cricket, one can see it in the bungalows and hotels and British influenced political institutions, in the 18th century temples and palaces, in the statues to contemporary politicians and in the store windows displaying video machines. One can see in the city of Kandy a struggle by a people to find an identity for themselves which has dignity which makes them a participant in the future and yet retains ties to the past.

In conclusion, we have examined not only the history of a place but the history of a political paradigm in that place as inscribed in the landscape. The Sinhalese Buddhist political paradigm was clearly worked out in pre-colonial Kandy. It was concretized in the landscape of the place. The British were never able to make this political paradigm mesh with their own and as a result they failed to achieve the political legitimacy that they sought. They did, however, achieve a degree of hegemony over the Sinhalese elites. This was achieved both consciously and unselfconsciously through landscape modification. And yet, as I have shown, the romantic landscape model unconsciously contained the denial of that hegemony in the form of a revival of pre-colonial Sinhalese landscape elements. Although the British tried to depoliticize Kandy, they were only able to mask not eliminate its political content, for the landscape models used were inherently political. The present government has attempted to achieve legitimacy by meshing a European model of democracy with a revived version of the Kandyan paradigm. But it is a curious *mélange*, drawing upon the language and environmental symbolism of absolute monarchy in a democracy, and of Buddhism as a *de facto* state religion in a multi-ethnic society with a highly politicized Tamil minority demanding separation.

And yet in spite of its incongruity, the model has been effective. According to Evers (1972, p. 15) there is still a longing among the Sinhalese for a king-like figure – someone who has charisma, who can, as Clifford Geertz (1983) says, touch the animating center of the society – a center which is profoundly Buddhist. Geographically, the symbolic center of the society is located in Kandy. It is built into the landscape of the place, the Temple of the Relic, of the gods, and the palace of the kings. The president touches that animating center when he stands among the sacred objects in that place and addresses his countrymen, when he tells them that they must create "a new society based upon the teachings of the compassionate one" and that "it is our duty to safeguard the Buddha Sasana" (Jayawardena 1974, p. 131). It is here in the Temple of the Relic in Kandy that the president for a few brief moments manages to link the present to a pristine pre-colonial political order. It is here in this place that the indignities of British rule are momentarily forgotten and the present troubles of a multiethnic state ignored. It is here that the Buddhist utopian vision is made to seem, for an evanescent moment, plausible.

Acknowledgements

My research in Sri Lanka was funded by the Social Sciences and Humanities Research Council of Canada in 1983 and by the Travel Fund of the University of British Columbia in 1985. I am grateful to Professor Gerald Pieris, Head of the Department of Geography, University of Peradeniya, who invited me to be a visiting research fellow in 1983. I am indebted to my field assistants Mr Shanta Hennayake and Mrs. Nalini Hennayake and to Nancy Duncan, John Agnew and Shanta Hennayake for comments on earlier drafts of this paper.

References

Bechert, H. 1978. S.W.R.D. Bandaranaike and the legitimation of power through Buddhist ideals. In *Religion and legitimation of power in Sri Lanka*, B. L. Smith (ed.), pp. 199–211. Chambersburg, PA: Anima Books.

Bennett, J. W. 1843. *Ceylon and its capabilities.* London: W. H. Allen.

Bingham, P. M. 1921. *History of the Public Works Department, Ceylon, 1796 to 1913.* Vol. 2. Colombo: H. R. Cottle.

Boyd, W. 1889. Autobiography of a Periya Durai. In *Ceylon Literary Register.* Vol. 3, Aug. 1888–July 1889. Colombo: A. M. & J. Ferguson.

Cave, H. 1912. *The book of Ceylon.* London: Cassell.

Culavamsa, Part 1 1953. W. Geiger, trans. Colombo: Ceylon Government Information Department.

Culavamsa, Part 2 1953. W. Geiger, trans. Colombo: Ceylon Government Information Department.

Davy, John 1821. *An account of the interior of Ceylon and of its inhabitants with travels in that island.* London: Longman, Hurst, Rees, Orme and Brown.

Dimmitt, C. & J.A.B. Van Buitenen (eds & trans.) 1978. *Classical Hindu mythology: a reader in the Sanskrit Puranas.* Philadelphia: Temple University Press.

Dolapihilla, P. 1959. *In the days of Sri Wickramarajasingha, last king of Kandy.* Maharagama: Saman Press.

Dougherty, Rev. J. A. 1890. *The East Indies station or the cruise of H.M.S. Garnet, 1887-90.* Malta: Muscat Printing Office.

D'Oyly, J. 1917. *Diary of Mr. John D'Oyly.* H. W. Codrington (ed.), *Ceylon Branch Royal Asiatic Society.* **25**, (69). (Colombo: Colombo Apothecaries Co. {Original 1825}).

Evers, H-D. 1972. *Monks, priests and peasants.* Leiden: E. J. Brill.

Farrer, R. 1908. *In Old Ceylon*. London: Edward Arnold.

Ferguson, A. M. No Date. *Map of the hill country of Ceylon showing the positions of the principal coffee estates*. Colombo: A. M. Ferguson.

Forbes, Major 1840. *Eleven years in Ceylon*. Vol. 2. London: Richard Bentley.

Geertz, C. 1973. *The interpretation of cultures*. New York: Basic Books.

Geertz, C. 1980. *Negara: the theater state in 19th century Bali*. Princeton: Princeton University Press.

Geertz, C. 1983. Centers, kings, and charisma: reflections on the symbolics of power. In *Local knowledge: further essays in interpretive anthropology*, C. Geertz (ed.) pp. 121-46. New York: Basic Books.

Godakumbura, C. E. 1961. *Sinhalese literature*. Colombo: Colombo Apothecaries.

Gramsci, A. 1971. *Prison notebooks*. London: Lawrence & Wishart.

Hurst, J. F. 1890. The enchanted road to Kandy. In *Images of Sri Lanka through American eyes: travellers in Ceylon in the 19th. and 20th. centuries*, H. A. I. Goonetileko (ed.) pp. 198-205. Colombo: International Communication Agency, United States Embassy, 1976.

Jayawardena, J.R. 1974. *Selected speeches and writings*. Colombo: H. W. Cave & Co.

Merton, T. 1968. *The Asian Journal of Thomas Merton*, N. Burton, P. Hart & J. Laughlin (eds). New York: New Directions Books.

Obeyesekere, G. 1972. Religious symbolism and political change in Ceylon. In *The two wheels of the Dhamma*, B. L. Smith (ed.). AAR Monograph 3. Chambersburg, PA: American Academy of Religion.

Obeyesekere, G. 1984. *The cult of the Goddess Pattini*. Chicago: University of Chicago Press.

Phadnis, U. 1976. *Religion and politics in Sri Lanka*. New Delhi: Manohar.

Pridham, C. 1849. *An historical, political and statistical account of Ceylon and its dependencies*. Vol. 1. London: T. W. Boone.

Rogers, K. R. 1903. I am more and more delighted with this island and its people. In *Images of Sri Lanka through American eyes: travellers in Ceylon in the 19th and 20th centuries*, H. A. I. Goonetileke (ed.) pp. 244-55. Colombo: International Communication Agency, United States Embassy, 1976.

Seneviratne, H. L. 1978. *Rituals of the Kandyan Stater*. Cambridge: Cambridge University Press.

Seneviratne, H.L. 1984. Continuity of civil religion in Sri Lanka. *Religion*, **14**, pp. 1-14.

Sirr, H. C. 1850. *Ceylon*. Vol. 1. London: William Sholberl.

Stewart, J. 1862. *Notes on Ceylon and its affairs during a period of thirty eight years, ending in 1855*. London: Privately Published.

Tambiah, S. J. 1986. *Sri Lanka: ethnic fratricide and the dismantling of democracy*. Chicago: University of Chicago Press.

Williams, R. (1973). *The country and the city*. London: Chatto & Windus.

Wright, T. Y. 1951. *Ceylon in my time, 1889-1949*. Colombo: Colombo Apothecaries.

12

Beijing and the power of place in modern China

MARWYN S. SAMUELS
& CARMENCITA M. SAMUELS

While modernizing, China should be careful to preserve as much of its valuable past as possible. . . .At one time, many people held that a new Beijing could be built only when the old Beijing was torn down. After removing many old buildings, including its city walls, Beijing is now much more modern. But many people have come to realize that Beijing is no longer "Beijing."

Beijing Ribao, 22 February, 1986

It is only fitting that a chapter on the landscape consequences of China's historic encounter with the West and with 20th century revolution be included in a volume dedicated to the memory of David Sopher. Born of Sephardic lineages arrived on the China coast and raised in Shanghai, the veritable cauldron of modern Chinese revolution, David Sopher was himself witness and party to the many opportunities, as well as the many anxieties of Chinese places caught between the press of Confucian-inspired tradition and the impress of Western-inspired revolution. Here, as elsewhere, he was no stranger to the tribulations and triumphs of places and peoples caught in the margins between the powers of a compelling past and the demands of a revolutionary future. Hence, as we explore some of the ambiguities and meanings of historic marginality on the land and in the built environment of China, David Sopher – the individual, the cultural geographer and the man reared on Chinese shores – is not far from our thoughts.

The problem

The general issue of historic marginality in modern China, or the question of whether and how Chinese traditions might remain salient in an era of revolutionary change is of long duration. Beginning with the late 19th century Confucian reform movement and its slogan "Chinese Learning for Essence, Western Learning for Use," and culminating more recently in efforts to "Sinify Marxism" or to define "Socialism with Chinese characteristics," the issue has been in the words of Tu Weiming, "the central Problematik of contemporary China" (Tu 1979, p.22). Indeed, amidst the storms of revolutionary paroxysm that swept across China in the 20th century, a great war between "ancients and moderns," conservatives and iconoclasts, romantics and revolutionaries has been fought almost without respite for the better part of the last 150 years. For the most part, no doubt, that war has been fought, just as it has also mainly been studied, on the grounds of ideological rectitude, and over the kinds of political, economic and social institutions appropriate to a modern Chinese nation-state, both as Republic and as People's Republic (Levenson 1965, Wakeman 1973, Cohen 1984). However, at the same time, so too has it been a war fought quite literally *on the ground*, in the built environment, in the forms and structures that occupy the land, and over the salience of particular places.

Figuratively and literally, the landscape of modern China has been a great battlefield in which the issue of whether and how to use, change, preserve, adapt or take inspiration from the landscape heritage of Confucian China has been the geographical and architectural counterpart to the struggle over the intellectual and institutional salience of the past.

Figuratively and literally too, the result of that struggle has been a cultural geography almost everywhere touched by ambiguity – a geography filled with places signifying change, yet also a geography of places that have somehow compelled loyalty and defied change. Ironically too, they are often the same places now caught somewhere in the margins between past and future, tradition and change. And, of all these places, none has been more prominent or more paradoxical in its visage than the City of Beijing.

Having served as the "Celestial Capital" of Imperial China, the formal *axis mundi* of the imperial cult and secular metropole of the Confucian world order for more than 600 years, Beijing has few equals in China either in terms of its historical significance, or in terms of the sheer number and volume of historic sites, buildings and artifacts reminiscent of imperial traditions. Of all places in China, Beijing is the place *par excellence* of remembrance, if only because its landscape is everywhere littered with reminders of past greatness. However, since 1949 the city has also served as the central repository of Socialist China, the administrative core and symbolic hub of revolution. The landscape of modern Beijing is equally and everywhere littered with the record of modern social, political and economic change. Not coincidentally, these two Beijings – traditional Confucian and modern Socialist – have confronted one another in ways that have had few equals elsewhere in China. Hence it is that we turn our attentions here to the power of the place, Beijing, and to that confrontation which here, perhaps more than anywhere else, summarizes the larger stakes in the battle over the landscape of modern China.

Celestial Beijing

With the exception of the first 50 years of the Ming Dynasty (1368–1644), from 1270 A.D. when Kublai Khan selected it as the site for the capital of the

Yuan (Mongol) Empire to February 1912 when the last emperor abdicated, Beijing was the principal seat of emperors and the cosmo-magical core of Imperial China. Rebuilt in 1420 by the third emperor of the Ming Dynasty, Beijing was the unequaled central place of the Central Kingdom, the hub of the Sinocentric world order, the core of its Confucian bureaucracy, the scholar–official elite, and the center of power. Other places in the Chinese Empire might have been more scenic, more exotic, more sophisticated, more exposed to the outside world, more inventive in the arts and sciences, and certainly more comfortable. But, no place in China was more powerful, more thoroughly imbued with the symbols of cosmic and earthly authority, or more given over to the business of statecraft. Similarly, in a society whose elites were trained as scholars, gained status-rank and position mainly by civil service examination, served as the principal purveyors of orthodox ideology, and were schooled in the arts of antiquarianism, no place was more the seat of learning and tradition, or more inherently conservative than Imperial Beijing.

As Paul Wheatley, Arthur Wright and others have already explored the depths of the design cosmology of imperial capitals in China, we need not here repeat their powerful and extensive arguments (Chang 1970, 1977, Wheatley 1971, Wright 1977). Suffice to say that the overall design of Imperial Beijing closely followed two basic and overlapping liturgies: (a) the cult of the emperor, the cosmology of which defined him as the "Son of Heaven" and his place as the *axis mundi*, the core out from which all earthly virtue and power extended, and (b) an arcane, partly shamanist, partly Daoist and later Confucian-assimilated orthodox geomancy (*feng-shui*) based on such ritual and divination texts as the *Zhou Li* ("The Rites of the Zhou Empire") and the *Yi Jing* ("Book of Changes"). These two traditions combined to establish the basic design principles of centrality, axiality and rectilinearity so characteristic of all traditional Chinese capitals, including Beijing. Indeed, they served, alongside craft traditions, to establish the rules of geometric modularity for that part of the landscape that served to symbolize and sanctify the Confucian cosmology and social order (Samuels, C. 1986). In this respect, the Confucian city, like the Confucian house and Confucian society, was highly regimented. Its layout and structure epitomized by a seeming endless maze of walled compounds within walled compounds within walled compounds were imbued with the signs of power, authority, and hierarchy, and nowhere more so than in the austere formality of Imperial Beijing (Wu Liangyong 1986, pp. 54–74).

To be sure, deviations from the ideal plan were also accommodated in practice. Although rectilinear, the layout of Beijing during the Ming and Qing periods was not the perfect square, divided into nine equal units as decreed by the orthodox prescriptions of the *Zhou Li*. The northwest corner of the outer city wall was turned on an angle in order to better accommodate the principal water supply system for the city and its moats. Moreover, in time, as the city grew the area immediately south of the southern wall of the Imperial City was filled in, separately walled, and eventually incorporated into the greater city of Beijing, but at a scale far different from that of the northern districts. During the Qing Dynasty

(1644–1911), the Manzhu government sought to formalize this distinction by prohibiting ethnic Chinese from living in the northern district. And, while the effort never fully succeeded, it was nonetheless immortalized on many Western maps as a distinction between the northern "Tatar City" and the southern "Chinese City" of Beijing.

Deviations from the ideal arose as well from the mundane functions of the imperial metropole. Beijing was, after all, not only the cosmo-magical core. It was also the business hub of empire-wide communications, the principal residence of scholar-official elites as well as their many minions, and the place *par excellence* for the congregation of the influ-ential and the influenced. Whatever its celestial pretensions, the practical result of many people of all kinds and substance living within the intra-mural confines of the Imperial City was an often unkempt pattern of small alleyways or lanes (*hutong*) filled with an irregular maze of all types and all scales of courtyard compounds (Samuels, C. 1986). Furthermore, interspersed in the residential landscape of Beijing were all manner of temples, monasteries, schools, restaurants, warehouses, workshops, and markets. In time too, whole new districts emerged to lend character to the cityscape. One of the most famous of these was the Liulichang district in the southern city. Here, in the small lanes of Liulichang, a district arose mainly to service the needs of scholar-official elites. Filled with shops selling the tools of the scholar's trade – ancient and modern printed books and manuscripts, ink stones, ink, brushes, brush holders, paper weights, artist supplies, art, antiquities and the like – Luilichang became one of the most famous haunts of intellectuals and officials in the Empire. However, Liulichang was not part of the original plan of the city. Rather, it had been a kiln site located outside the southern city walls that served to manufacture the colored tiles used to decorate the roofs of the Ming imperial palaces during their construction in 1420 (Yee 1984).

Other secular patterns might also be identified, but the point here is simply to note that Imperial Beijing was a place where the profane met the sacred in ways that made for informal deviations from the ideal cosmological plan and layout. However, this is not to suggest that such secular patterns challenged the celestial definition of the city. On the contrary, traditional Beijing was thoroughly dominated by the official landscape. Indeed, the design prescriptions of the celestial city caused all manner of nuisance to the mundane flows of urban life. For example, no commoner was permitted to enter the Imperial City via the large, centermost doorway of the "Front Gate" (*Qianmen*), as this was the ex-clusive preserve of the emperor and was opened only on ceremonial occasions. Two smaller side doors in the gate were available for plebian traffic. However, as this was the main entryway into the city, the conges-tion could at times be as intolerable as any modern traffic-jam. Similarly, the roadway that served to delineate the sacred north–south axis lead-ing from the Forbidden City to the Front Gate of the city was itself taboo territory to all but the emperor and his official retinue. It effectively divided the northern Imperial City into two halves – west and east – neither of which was directly accessible to the other. Commoners seeking to cross this divide were required – at the pain of execution – to go entirely around the city to get to the other side.

In short, whatever secular deviations may have emerged on the landscape of Imperial Beijing, the city as a whole, its layout and its built environment was nonetheless conditioned by the prescriptions of the imperial cosmology and its geomantic modularity. The city's principal thoroughfares and buildings were so sacred that one approached in awe, not to mention fear, careful to observe every nuance of ritual aimed to affirm and proclaim one's fealty and reverence for the Harmony and Virtue of the Imperial and Confucian Order. Failure to observe such rituals could result in all manner of punishments, and the construction of buildings that deviated from prescribed norms could result in their confiscation and demolition, not to mention the banishment of their owners. Therefore, whatever concessions may have been made to practicality, the driving force in the design, layout and construction of Imperial Beijing was its role as the "celestial capital."

Desanctification

However, some 600 years after the city was first designed and some 440 years after its reconstruction by the Ming, the two Beijings – celestial capital and mundane metropole – met in open conflict, the celestial shifting ground to accommodate a host of new and alien prerogatives and demands which, in effect, destroyed the once sacred domain of Imperial Beijing. In less than 90 years, from 1860 to 1949, the whole fabric of Confucian China came apart at the seams and, along the way, the cosmology that had so long dominated the landscape of the celestial city fell into abandonment. Indeed, on 12 February, 1912, with the formal abdication of the last emperor of China, the landscape of Beijing was officially desanctified.

If any one event can be said to have launched the process leading to the desanctification of Beijing, it was most assuredly the joint Anglo-French destruction of the Yuanmingyuan Summer Palace on 18 October, 1860. Located in the northwestern suburbs of Imperial Beijing, the Yuanming-yuan comprised a vast complex of some 40 major palaces, gardens and man-made lakes, the design and construction of which had begun in 1709 under the Kangxi emperor. Completed during the period 1736–1745 under the guidance of the Qianlong emperor, it had stood for more than a century as the favorite personal residence of the Son of Heaven, and as one of the best known symbols of imperial elegance and power (Liu Dunzhen 1981, Wang Zhili 1981, Bai Rixin 1983). Its destruction in 1860, in short, quite literally brought the so-called Opium War and the impact of the West into the bedroom of the celestial emperor himself.

While the details of the attack on the Yuanmingyuan need not detain us here, its destruction in 1860 underscores much of the process that was to undermine and a half-century later bring about the collapse of the Chinese imperial system and its Confucian rationale. Most importantly, it constituted the first direct assault by the Western powers on the sacred landscape of Beijing, an event whose sequel occurred when, in 1900, the "Eight-Nation Army" arrived in Beijing to lift the Boxer seige of the foreign legation quarter, and proceeded to destroy various parts of Imperial Beijing itself including the sacred square in front of the Forbidden City and its Gate of Heavenly Peace. The "Thousand-Step Gallery" that had lined both sides of the square was burned to ashes, and the "Front Gate" to the Imperial City was burned to the ground, its outer battlements being reduced to rubble. In addition, like the Yuanmingyuan 40 years earlier, various other imperial palaces and properties were damaged and looted in the wake of the attack.

In one sense, of course, the destruction wrought by the Western powers undermined the sacerdotal character of Imperial Beijing. But, it was not merely in such military terms alone that they launched the desanctification of the landscape. After all, Beijing had been conquered before only to recover and be rebuilt according to traditional design principles. However, in this case the invading armies and their camp followers – the merchants, missionaries, and intellectuals – carried something other than ammunition in their packs. They brought along a baggage filled with a host of entrepreneurial, technical and progressivist ideas about urban life and industrial prosperity, as well as a body of political, social, and economic values which, once released, confronted the integrity of the Imperial–Confucian world system and its cosmology with a series of powerful secular threats in the form of modernization, Westernization, and revolutionary nationalism.

Suffice to say here that the first three decades of the 20th century witnessed the wholesale transformation of Beijing's urban landscape, a process whereby the secular and the profane ultimately overran the sacred and the celestial fabric of old Beijing. There are many examples of this process, but we need here only touch upon a few. For example, among the more prominent features of the Western impact on the landscape of Beijing we can cite the expansion of the Legation Quarter just to the southeast of the Forbidden City, the growth of the Wangfujing-Dongdan commercial core just to the east of the Forbidden City, the opening of new east–west arterials, the opening of the rail line along the southern wall of the Imperial City linking Beijing with the rest of China, the growth of the university district in the northwestern suburbs, and the growth of the Tianqiao "Red Light" district. By 1907, for example, the "Gran Hotel de Pekin" had been built on East Changan Avenue, and by 1928 what had originally been a small, dirt-packed lane leading to several princely estates, Wangfujing Lane, had become a tree-lined asphalt-paved boulevard, and the "Fifth Avenue" of modern Beijing (Wang Zhishu *et al.* 1983, Zong Quanchao 1985). By the early 1930s the construction of modern hospitals, office buildings, hotels, schools, colleges, theaters, shops, department stores, workshops, factories, public parks, sewage systems, trolley lines, electric and telephone lines, and all manner of the paraphernalia of a "modern city" emerged to confront the celestial city of Beijing with a profoundly secular threat (Gamble & Burgess 1921, *Jiudu wenwulue* 1935). In short, the mundane metropole increasingly acted to dominate the cityscape and to undermine the once celestial capital.

Indeed, if such structural changes served to underscore the secularization of the celestial city, its formal desanctification came in the wake of a new and revolutionary nationalism by which the cosmology of Confucian China, the cult of the emperor, and the arcane geomancy of the celestial city was finally destroyed. Simply stated, the formal desanctification of Imperial Beijing came abruptly and, in Confucian terms, on the singularly astounding afternoon of 12 February, 1912 when the last emperor of the last dynasty of Confucian China abdicated in favor of the Republic of China. On that date, 268 years after the founding of the Qing Dynasty, the two-millennia-old Imperial system came to its ritual end. In less than 15 years, the abdication of the child-emperor Xuantong (better known by his personal name, Pu Yi), his forced eviction from the Forbidden City in 1924, and the formal removal of the capital of Republican China to the City of Nanjing in 1927 served notice that Imperial Beijing and its once celestial imperial properties were now and, apparently forever, desanctified.

In the wake of Republican anti-Manzhu and anti-monarchial sentiments, the maintenance and use of the formerly sacred properties of the "Son of Heaven" were issues intimately tied to the precarious status of the vestigial monarchy itself. How to handle the person and family of the then seven-year-old emperor upon his formal abdication in 1912, as well as how to deal with his and their properties was to become the subject of much intrigue and manipulation on the part of erstwhile Republicans, revolutionaries, warlords, monarchists, and foreigners for the next four decades (Johnston 1934, pp. 160–79, Aisin-Gioro Pu Yi 1964, pp. 33–47, Aisin-Gioro Pu Jie 1985). By February 1912, for example, the southwestern corner of the once utterly taboo Forbidden City had been confiscated and converted into a residential and office complex for Yuan Shikai, the first president of the Republic of China. Similarly, according to the Articles of Abdication, the child-emperor Pu Yi and his immediate family were supposed to vacate the premises of the Forbidden City and remove themselves to the Yiheyuan Summer Palace. In fact, they remained in the Forbidden City until 1924 and, in the meantime, the royal family opened the Yiheyuan Summer Palace in 1914 as a public park, charging a nominal entrance fee for its maintenance.

Other members of the royal family were less fortunate. Many of their estates were quickly confiscated for public and private uses. In 1914, for example, the West City District estate of Prince Chun, the father of the Guangxu emperor, and the grandfather of Pu Yi, was taken over to serve as China University and later as Republican University. (It is now the site of the Central Conservatory of Music.) Furthermore, in late 1923, the new government together with representatives of the Qing royal house formed a so-called "Committee for the Readjustment of the Qing," the main purposes of which were to take inventory of all imperial properties and to determine which were to become "public" and which were to remain in "private hands." The former royal house had, as it were, truly fallen on hard times, being reduced from their once celestial status to the now thoroughly profane role of mere "private" citizens. And, to cap their fate, in 1924 the celestial emperor fled the Forbidden City on the heels of threatened assassination, leaving his Dragon Throne and formerly sacred palaces to become the now merely secular property of the state in the name of the "Palace Museum." In short, celestial Beijing was abandoned. To make its fate more complete, in 1927 the capital of Republican China was officially transferred to Nanjing.

This is not to say, of course, that there was no opposition to the desanctification of Beijing or to the collapse of the traditional order. On the contrary, during the first three decades of the 20th century various and sundry attempts were made not only to restore Beijing to its position as celestial capital, but also to salvage the Confucian tradition. Chief among these was the attempt by the arch-traitor Yuan Shikai, the former chief of staff of the Qing Imperial Army who, having betrayed his royal patrons to become the first president of the Republic of China, soon sought to betray his Republican patrons as well in an attempt to restore the monarchy, appointing himself emperor. To be sure, Yuan Shikai failed, and the conservative opposition was ultimately powerlesss to arrest the tide of secular nationalist demands. Indeed, in one sense, such conservative opposition served mainly to underscore the overall erosion of Confucian values in modern China. As Lu Xun (1980, vol. 4, pp. 188–7) put the matter in 1935:

> Since the start of the 20th century Confucius has had a run of bad luck, but by the time of Yuan Shikai he was once more remembered: not only were his sacrifices restored, but bizarre new costumes were designed for those offering sacrifice. This was followed by the attempt to restore the monarchy. That door did not open, however, and Yuan Shikai died outside it. That left the Northern Warlords who, once they felt their end approaching, also used Confucius as a brick to knock at the door of happiness. . . but since times had changed, they all failed utterly. Apart from failing themselves, they involved Confucius, making his position still more lamentable.

While this statement reflects Lu Xun's characteristic cynicism, it also underscores the lamentable stature and ersatz character of Confucianism by the mid-1930s. Indeed, it was a comment less in criticism of Yuan Shikai, than in judgment of the "New Life Movement" of 1934 and other Guomindang Party efforts to rescue an already vestigial Confucianism as a means to insulate and differentiate itself from the greater iconoclasm of Communists and other competitors on the left (Levenson 1965, pp. 105–8, Hsu 1970, p. 668). At the same time, however, it was also a comment remarkably insensitive to another, more complicated manifestation of the nationalist urge to rediscover and reassert intellectual, cultural and geographical continuity with the Confucian past. After all, just beneath the surface of Republican anti-monarchial and nationalist sentiments was a powerful sea of historical-geographic, cultural, and ethnocentric reference points necessary to the definition of the "the people" and "the nation" (Levenson 1965). Indeed, in light of the anti-imperialist and anti-foreign impetus of Chinese nationalism in the early 20th century, where else but from the heritage of Confucian China were revolutionary nationalists to draw their inspiration in seeking to establish a distinctively Chinese polity?

Ironically, it was left to the icononclasts, revolutionaries and Westernizers among the intellectuals and policy makers first of Republican China and later of Socialist China to salvage the heritage of Confucian China and its landscape. They did so, of course, not as advocates of the *ancien regime* or to restore Confucian China and its celestial cosmology, but in search of cultural salience, and in the secular effort to preserve and draw inspiration from the one inheritance that was, after all, uniquely their own. They did so as parties to what began in conjunction with the desanctification of the landscape as a Chinese "renaissance."

Renaissance ambiguities

One of the more ironic-yet-necessary corollaries to the rise of revolutionary nationalism in the early 20th century was a reborn interest in and concern for the heritage of China that came, for example, with the growth of a new historicism, fueled in large measure by Western teleologies and methods, that sought not only to reinterpret, but also to recover and preserve the Confucian past for modern China. Similarly, the "Baihua (Vernacular Language) Movement" of Hu Shih, and even the "New Culture Movement" of Chen Duxiu and other Marxists were almost as conservative as revolutionary in their efforts to adapt the past to serve the future of China.

In much the same light, aided by the birth of modern Chinese archaeology, and abetted by the growth of such fields as historical geography and architectural history, the landscape endowment of Confucian China became the subject of much scholarly research and the target for all manner of preservationist activity. Hence, for example, during the first three decades of the century Liang Sicheng and others launched intensive efforts to reconstruct the long-since lost or barely comprehensible medieval literature of imperial design and construction technique, the hallmark of which was Liang's work on the 11th century *Yinqzao fashi* or "Treatise on Architectural Methods" (Liang Sicheng 1983, 1984). Furthermore, supported by the growth of modern architecture, engineering and city planning, a new generation of Chinese architects and planners sought to recapture the spirit of the Confucian heritage in their search for "design authenticity." In this too, as in many other aspects of the early 20th century Chinese renaissance, the concern for historic preservation and design authenticity on the land arose ironically, yet also necessarily in response to foreign influences. It came, as it were, on the heels of similar sensitivities in late Victorian England, and especially during the inter-war period in Europe and America when the past became an exotic "foreign country" traversed by all manner of latter-day humanists and romantics (Lowenthal 1985). It came too in the form of influences derived from Great Britain when the British government promulgated historic preservation legislation in the form of the "Ancient Monuments Act of 1913." Closer to home, the public concern for historic preservation came to China as well from Japan when in

1919 the latter promulgated its "Preservation of Cultural Properties Act," the first such legislation in Asia and the precedent for similar legislative action in Republican China (Dunfield & Graham 1986, p. 42). Similarly, it came as well in the wake of the work of such Western architects and planners as Henry K. Murphey who, in obtaining rich commissions for the design of many new university campuses, libraries, office buildings and other sites during the first two decades of this century, sought to blend modern building materials and construction techniques with traditional Chinese designs (Cody 1986).

Moreover, from the early 1920s to the late 1940s the return to China of the first and second waves of Chinese architects and planners trained in Britain, Germany, the United States and Japan brought additional impetus to the search for design authenticity and preservation (Su Gin-djih 1964, Wu Guangzu 1983). Those like Lu Yenzhi (graduated from Cornell in 1924), Liang Sicheng (graduated from Pennsylvania in 1928), Chen Zhanxiang (graduated from Liverpool and London in 1938 and 1945), and many others were instrumental in the effort to restore the record, preserve the architectural endowment, and establish the conceptual ground for the adaptation of the traditional design heritage.

Along the way, the once celestial became the now secular and preserved stuff of museums, libraries, universities, protected historic sites and tourist attractions. Hence it was, for example, in 1927 that Lu Yenzhi won the national competition for the design of the Dr Sun Yatsen Mausoleum with a plan that adapted traditional design concepts to modern concrete construction methods, and to such Republican sensitivities as use of blue rather than imperial yellow tile. Hence too in 1935 the municipal government of Beijing sponsored and published an extensive historical-geographic guide to the "relics" of old Beijing, *Jiudu wenwulue* (Outline of the Cultural Relics of the Old Capital"). Similarly, it was during this period that much sub-mural local history came to figure prominently in Beijing and elsewhere, as did extensive studies on the histories and design philosophies of particular buildings, gardens, and estates (Hong Yeh 1981, pp. 39–40, 95, 107–111, Xu Zhaokui 1981, Hou Renzhi 1984a,b,c, Wang Canchi 1985, pp. 278–80). Indeed, not the least of the sites toward which such attention focused was the place where the process of Beijing's desanctification had begun, the Yuanmingyuan (Bai Rixin 1983, Wang Zhili 1983, Zhaoi Guanghua 1984, Wang Canchi 1985, pp. 282–7).

In short, the Republican period was as much filled by the impetus to cultural renaissance as it was schooled in the iconoclasm of revolutionary nationalism. Yet, if the two were mutually interdependent, so too were they potentially antagonistic. After all, in an era of revolutionary rhetoric and violence, the subtle distinctions between the scholarly pursuit of cultural salience and the conservative inclinations of modern nationalism were easily blurred or rendered moot. An entire generation of "radicals" and "revolutionaries" were soon to find themselves labeled "reactionaries" and "counter-revolutionaries," just as the progeny of a new secular inter-nationalism in the name of Marxism were soon to find themselves the principal inheritors of a highly culture-specific Chinese nationalism. And, just as such rectifications of names became increasingly muddled, so too did the issue of historic preservation and design authenticity on the land become increasingly ambiguous. Indeed, it became extraordinarily paradoxical once possession of the landscape endowment of Confucian China passed into the hands of the Communist Party and the People's Republic of China on 1 October, 1949 – coincidentally only a few days short of the eighty-ninth anniversary of the event that had launched the desanctification of celestial China and Beijing, the destruction of the Yuanmingyuan.

Socialist iconoclasm

In one of his many allusions to the modern fate of Confucian China, Lu Xun once employed the analogy of a large old house inherited by a poor young boy. The boy, Lu Xun suggested, could assume one of three principal attitudes towards his new found fortune. On the one hand, if he "feared the former owner of the house, he might be afraid to damage anything left behind, and hesitate, not even daring to enter the gate of the house." This, Lu Xun noted, would be "cowardly." On the other hand, the poor boy might instead "curse excitedly, and take a match to torch the place, thinking that in doing so he preserved his own integrity." This, Lu Xun thought, would be "muddleheaded." Finally, he noted, "not only because he envied the possessions of the former owner, but also because he now possessed them, (the boy) joyfully slipped into the bedroom of the house and fully availed himself of opium." This, Lu Xun complained, "is obviously a true waste" (Zhang Jingxian 1981, pp. 107–8.)

Taking his lead from Lu Xun's allegory, one recent critic, Zhang Jingxian, reflected further on the fate of that house, and the history of modern Chinese architecture and planning since 1949.
As he summarized it (1981, p. 108):

> The pity is that after liberation we also acquired the curse of this mistaken attitude. During the 1950s in our adoption of traditional designs we "joyfully" indulged in the indiscriminate waste of capital by the excessive use of colored tiles, the use of reinforced concrete to make *douqong*, and other fake decorations. Afterwards, a period of coercion set in during which labels were attached to everything. In "coming down with a big stick" and "seizing on mistakes," we experienced "hesitating, not even daring to enter the gate," afraid of being stained by the forces of " feudalism," "capitalism," and "revisionism". Then especially during the ten years of calamity (i.e., the Cultural revolution, 1966–1976) when the Lin Biao and Jiang Qing anti-revolutionary gang took the relics of classical architecture among the "Four Olds" to be swept away, we "took a match to torch the place" or thoroughly smashed it. Thus, according only to preliminary data, among the 77 national and municipal protected

cultural sites in Beijing, some 44 were destroyed; of 114 large, medium and small gardens in the City of Suzhou, at least 38 were completely destroyed. . . .

We do not here need to follow Zhang through the rest of his depressing recitation of destroyed or damaged historical sites to be reminded that since the founding of the People's Republic of China, the war over the landscape has taken many tortured twists and several violent turns. Should additional evidence be required, we need only cite the official statistics for the City of Beijing, according to which it is reported, for example, that in addition to the demolition of the celestial city's once massive walls and various other historic sites during the Cultural Revolution, the local steel mill melted down 117 tons of gold historical relics, and the local paper mill recycled over 320 tons of books and manuscripts, as well as some 185,000 scrolls of calligraphy, 2.357 million volumes of classical texts, and some 538,000 other items of historic value (*Xinxinxiangrong de Beijing* 1984, p. 123). In short, the landscape endowment and cultural heritage of Confucian China came under an assault the likes of which, ironically, can perhaps best be compared with the worst excesses of the Western powers and Japan in their own pillage of Confucian China. A hundred years separated the actions of the so-called "Gang of Four" led by Jiang Qing during the Cultural Revolution and those of Lord Elgin in 1860, but it was a short hundred years, linked more closely by the road to desanctification along which any number of burned, looted or deteriorated hulks stood as silent sentinels to the loss of their once celestial and geomantic powers.

For that matter, the obvious iconoclasm of the Cultural Revolution aside, the secular transformation of the urban landscape of Socialist Beijing since 1949, like its Republican counterpart, underscores the continuity of revolutionary desanctification that began, as we have seen, at least as early as 1900. In the wake of such campaigns as the 1950s effort to transform former "consumer cities" into modern "producer cities" the proletarian ethos of Socialist China generated a new urban landscape whose own sentinels supplanted the forms of Celestial China with the steel and reinforced concrete structures, and the steel and macadamized networks of an industrial age (Zhao Xiqing 1984, *Xinxinxiangrong de Beijing* 1984).

We need not here explore the details of that process, save to note that in the case of Beijing, for example, during the eight years from 1949 to 1957, the urban population of the city almost doubled from 1.76 million to 3.41 million, and its industrial labor force more than tripled from 320,000 in 1949 to 1.08 million in 1957. To serve the housing requirements of its fast swelling population, newly completed 4-to 8- storey concrete apartment blocks grew from a total of 4.79 million square meters in 1949 to 53.73 million square meters in 1957, more than a ten-fold increase which nevertheless only barely kept pace with the increasing pressure for new housing stock. Indeed, by 1980, continued population growth and the slow pace of housing construction during the Cultural Revolution conspired to produce an average per capita urban living space in Beijing of 4.79 square meters, a mere 0.04 square meter improvement over what it had been

30 years earlier in 1949 (*Xinxinxiangrong de Beijing* 1984, p. 51, 96, 360, Ma, 1981, p.231).

In short, as the urban economy of Beijing continued its transformation from an Imperial–Confucian service center towards an industrial production center, the once celestial built environment gave further ground to the now thoroughly secular landscape of a "socialist city." What had begun in the first three decades of the 20th century now escalated as whole districts of the former Imperial City were bulldozed and replaced, as whole new districts grew in suburbs now more readily linked to the inner city by the removal of its once massive external walls, and as the wall itself was replaced by an encircling, concrete and macadam "ring road." In the process too, historic *pailou* (memorial arches) and other traditional street furnishings were removed or demolished, whole districts of classic *sihevuan* (courtyard houses) were replaced by medium-to-high rise apartment and office structures, and any number of historic buildings and sites were taken over, remodeled, demolished, or simply allowed to deteriorate (Zhang Kaiji 1985a,b, 1986a,b,c). Such, it seems was the price to be paid for "modernization." After all, Beijing was no longer the Celestial Capital of Confucian scholar-official elites, but the new core of Socialist China and its proletarian (i.e. worker-peasant) and Communist Party elites.

And yet, if "socialist", Beijing nonetheless remained "Chinese." Indeed, it became the symbolic core of Chinese socialism which, if born of European and later of Soviet themes, was nevertheless a variant derived of Chinese sensitivities. Here too, in short, amidst the iconoclasm of socialist revolution, and even amidst the extraordinary iconoclasm of the Cultural Revolution and its attacks on all vestiges of Confucian China, the ambiguities of the early 20th century Chinese "renaissance" reemerged. The search for an impetus to cultural integrity remained undeniable and, if anything, those earlier ambiguities became the even more paradoxical, but powerful Chinese continuities of the People's Republic of China.

Preservation and paradox

Whatever else one may say about the record of iconoclasm in the People's Republic of China, it is at least a record not nearly as simple or as evil as polemicists on the right might wish (Leys 1976, 1986, Butterfield 1987). After all, the Chinese "renaissance" was as much the possession of Chinese socialists as it was the product of Republic nationalists. Indeed, it became the hallmark, if also the intellectual dilemma of left-wing artists and writers who, in seeking to transform China, nonetheless also sought to recover and hold on to that which was uniquely Chinese (Spence 1981). Moreover, it was also a search the ideological sanctions for which could be traced in the liturgy of the Chinese Communist Party (CCP) at least to 1942 when Mao Zedong gave his famous series of "Talks Delivered at the Yenan Forum on Artistic and Literary Work." Here, in his criticism of the "slavish imitation of foreign fashion," and in his command "to master our own national style," Mao Zedong provided the conceptual ammunition for what the intellectuals and ideologues, planners and politicians of Socialist China would seek in the name of the slogan "socialist in content, national in style" (Mao Zedong 1959).

Among the early beneficiaries of that sensitivity was the effort to protect and draw inspiration from the landscape endowment of Confucian China, especially in Beijing. During the winter of 1948–49, for example, as Republican Beijing prepared for the assault of Communist forces, and as the latter began to set their artillery sights to shell the walled city from the Western Hills, an event occurred that is perhaps especially revealing of that sensitivity. A CCP guerilla unit was apparently infiltrated into the city to meet with Professor Liang Sicheng in order to determine which historic sites and buildings ought to be spared an intended massive shelling of the city. Together, Liang and CCP officers selected points and zones on the map of the city that were to escape direct fire. Fortunately, as the city eventually surrendered without a fight, the selection became unnecessary, but the effort nonetheless revealed a concern for the protection of historic sites that would be reinforced once the city was taken over.

Indeed, within three days of the "liberation" of Beijing on 31 January 1949, the Education Department of the CCP Military Affairs Commission established a "Ministry of Culture" to protect and administer the city's many historic sites. Yin Da, the director of the army's Education Department was appointed "Minister of Culture," and a group of Education Department officers were individually appointed to oversee the administration of such sites as the Palace Museum, the Beijing Library, the Temple of Heaven and others. For example, on 11 February 1949 Luo Ge was officially placed in charge of the Palace Museum. Thirty-seven years later Luo would recall with some amusement how he arranged a formal meeting on 6 March, 1949 between Ma Heng, the Republican-appointed museum director (and a former professor whom Luo had known while a student at Peking University) and Yin Da. With some humor and much obvious irony, Luo Ge chose the former throne room of the Qing emperors, the Hall of Supreme Harmony, as the site for the meeting at which Yin Da presided. Moreover, in order to assure the continuity of the museum's curatorial and archival functions, it was at this meeting too that Professor Ma Heng and the then current staff were reconfirmed in their positions as on-site administrators and protectors of the Palace Museum (Luo Ge 1986, p.23, 43).

In short, at the very onset of the new regime, formal efforts were made to protect major historic sites. Moreover, by 1952 that intent was reconfirmed by the promulgation of the first PRC national legislation for the protection of historic relics and to prevent the export of so-called national treasures, a regulatory process that was further refined at the Beijing municipal level in 1957. Similarly, that concern was expanded upon at the national level in 1961 with promulgation by the State Council of a series of "Provisional Regulations on the Protection of Cultural Relics," and in 1963 by the "Provisional Measures for the Administration and Protection of Cultural Relics" issued by the Ministry of Culture. Indeed, during the first two decades of the People's Republic of China, concerted efforts were made to protect historic sites at virtually all local, provincial and national levels.

We need not here reproduce the many specific examples of historic buildings and sites selected for preservation during this period. Suffice it to say here that in the case of Beijing, in 1957 the municipal government identified some 39 sites, and in March 1961 the State Council of China officially designated 180 sites for preservation around the country, of which 18 were located in Beijing. According to official records, furthermore, during the 17 year period from 1 October 1949 to May 1966 (the beginning of the Cultural Revolution), the Beijing Municipal Department of Cultural Relics alone recorded and included under its adminstrative authority some 2,666 temples, 616 "ancient buildings," 700 tombs, 51 major stone steles and carvings, and an additional 129 "significant historic sites." Similarly, from 1949 to 1959 they had overseen the archaeological excavation of some 671 tombs, including the Ming Imperial Tombs excavated in 1956–58, and were responsible for the protection of some 43,621 separate relics taken from these tomb sites (*Xinxinxiangrong de Beijing* 1984, pp. 122–3).

Simply stated, the endeavor to protect historic buildings and sites was by no means inimical to the interests of the Chinese Communist Party or the new state. Much as in the case of the Soviet Union and other so-called revolutionary societies, whatever their ostensible iconoclasm, the historical powers of certain places remained virtually undeniable. Indeed, in the case of Beijing, even as the city underwent all manner of infrastructural and architectural transformations, in at least two respects the city defied change and reasserted its historic powers as a uniquely symbolic center. Simply stated, (a) the choice of Beijing as the capital of Socialist China, and (b) the choice and design of Tiananmen Square as the symbolic core of Beijing both served to underscore the paradoxical salience of history in modern China.

The selection of Beijing as the capital of Socialist China was itself, of course, a powerful statement on behalf of continuity. It was a choice made, after all, not without a range of options. The choice was made in August 1949, six months after the "liberation" of Beijing, and in the context of at least three other principal candidates: Xian, Loyang and Nanjing. While each of the latter, and Beijing as well, enjoyed certain geographical, strategic and logistical advantages, and while each also enjoyed the patronage of various individuals in the CCP leadership, the historic prominence of Beijing as the symbolic core of a united China was an undeniable incentive towards its choice as the capital of a newly united China.

Indeed, in an act that was intended to cement the symbolic mandate of the new regime in the popular consciousness, Mao Zedong and the leadership of the CCP chose to proclaim the existence of the new People's Republic while standing on the balcony of one of the most obvious symbols of the once celestial order, the "Gate of Heavenly Peace" (*Tiananmen*), principal gateway to the former Forbidden City. If, as Jonathan Spence has noted, the "Gate of Heavenly Peace" has borne "quiet witness to the new paradoxes" of modern Chinese history, it never did so more emphatically than on 1 October 1949 (Spence 1981, p. 17). While ostensibly aimed to commemorate the May 4th Movement, the leadership of the CCP knew full well that it was exactly from the spot on which they stood that for the previous two centuries the formal proclamation of all new imperial reign periods and titles, and the accession of all new emperors to the cosmo-magical Mandate of Heaven were publicly announced.

For that matter, the leadership of the CCP also went several steps further to confirm the thread of symbolic continuity. They not only launched the new People's Republic from the principal ceremonial gate of the once taboo Forbidden City, but also chose part of its inner domain as the site for their own private residential and office complex. Ironically, they followed the 1912 precedent of Yuan Shikai by taking over the southwestern corner of the palace–park complex of the Forbidden City known as Zhongnanhai (Middle and South Seas). Furthermore, in deciding upon a symbol appropriate to the official State Seal, they selected – of all things – a stylized version of the Gate of Heavenly Peace. They chose Liang Sicheng and his wife Lin Huiyin to design the seal in a manner that was intended to evoke the association of the operational core of New China with the operational and symbolic core of Old China.

But, it was not only in these choices that the leadership of the CCP revealed their intention to preserve and even to restore the symbolic powers of Beijing. They did so as well in considerations over the re-design and use of the inherited landscape. From 1949 to 1958, the senior architects and planners of the new regime, many of whom – like Liang Sicheng and Chen Zhanxiang – were carryovers from the former Republican period, renewed the search for design authenticity almost as if it had been uninterrupted. Like others, they did so in the context of efforts to build socialist content, while retaining national forms.

Perhaps naturally, those efforts gave rise to a number of design controversies, some of which centered around a curious form of indigenous chinoiserie that emphasized the use of such external furnishings as golden roof tiles and fake *dougong* attached to otherwise clearly modern structures. As Zhang Jingxian was to note much later, this was perhaps a case of the "joyful" but nonetheless "indiscriminate waste of capital." (Zhang Jingxian 1981, p. 108). However, throughout China, many of the design and planning controversies of the 1950s also centered around the issue of whether and how best to protect, use, change and gain inspiration from the inherited urban landscape (Wu Liangyong 1983). In the case of Beijing, the isssue focused primarily around the fate of its massive city walls, and the design or redesign of its symbolic and functional core.

Various attempts were made during the first few years of the PRC to adapt the inherited landscape of Beijing to the symbolic and functional needs of the new capital. However, two were perhaps most important. In the early 1950s Liang Sicheng proposed that the 39.7 kilometer – long outer wall of the former Imperial City be incorporated into the landscape of the "socialist city" as a public museum and "park." According to his plan, the top of the wall would become a public promenade, interspersed with gardens of potted plants and open-air restaurants. Similarly, many former guardhouses on the wall would be converted into galleries, museums, exhibition halls, shops and restaurants (Li Xiongfei 1985, p. 79). The once formidable barrier-wall of Celestial Beijing would become, as it were, a now open and welcoming urban gallery for the growing and increasingly prosperous population of Socialist Beijing. The wall would be preserved, yet also transformed in social meaning. It was, in short, a picture peculiarly appropriate to the early optimism of the new regime.

It was such optimism too that led Chen Zhanxiang and Liang Sicheng in February 1950 to put forward a bold new plan for the redesign of Beijing's symbolic core. They proposed an entirely new yet strikingly traditional design for the socialist capital that would, on the one hand, protect the old imperial core as an historic center of museums and parklands while, on the other hand, it would have shifted the symbolic and administrative core of Socialist Beijing to a site just beyond the western wall of the former Imperial City. Laid out on a traditional north-south axis almost parallel to the Forbidden City, their "New West City District" was to have been an extension of the traditional geometric grid, centered on a modular cluster of central government office buildings and the national legislature. Intended partly to further integrate the old city and its Palace Museum complex into the public domain, and to preserve links with the spatial symbolism of the old city, the design was primarily aimed to emphasize the role of Beijing as a new center commensurate with the birth of a new Socialist China (Chen Zhanxiang & Liang Sicheng 1950/1983, Liang Sicheng & Chen Zhanxiang 1986, pp. 1–31).

For that matter, Chen Zhanxiang, then senior planner for the City of Beijing, went several steps further to propose the more complete integration of the Palace Museum complex and the former imperial parklands of the

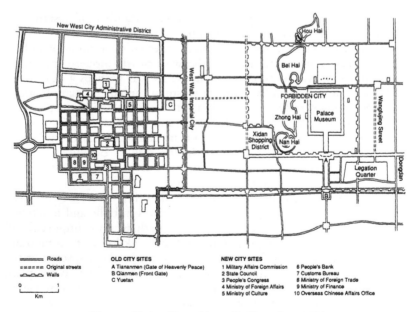

Figure 12.1   Chen-Liang Beijing plan, 1950.

Forbidden City into the profane flows of urban life. He proposed that the western wall of the Forbidden City be removed in order to open the entire length of the former imperial parklands and its lakes (Beihai, Zhonghai and Nanhai) into a vast municipal park. Bounded on the east by the Palace Museum complex and on the west by what he proposed to become a redeveloped Xidan commercial and residential district, this municipal park would then constitute a pivotal greenbelt running north-south along the geographical center of the new city. As this would necessitate the removal of the residential and office complex of the CCP leadership from the two southernmost lakes of Zhongnanhai, he proposed that they be relocated in the new West City District in what was to become Diaoyutai Park (the present State Guesthouse complex.)

Borrowing from but also taking license with design precedents in London, New York and other major world cities, Chen Zhanxiang's proposals called for a city built around two separate but intertwined themes: national unity and authority centered on a newly proclaimed West City District, and national culture and popular prosperity centered on a newly expanded Palace Museum Parkland and Xidan Commercial District. Chen Zhanxiang's new Socialist Beijing would have entailed, on the one hand, a popular core akin to New York's Central-Park–Metropolitan-Museum–Fifth-Avenue complex, and on the other hand, a political core loosely akin to the Capitol Hill-White House corridor of Washington, DC. Moreover, added to the overall master-plan put forward by Chen Zhan-xiang and Liang Sicheng, these two cores would each have expanded northwards, the one into a vast complex of Academy of Science research institutes, universities and colleges, and the other into the extended park-lands of Houhai, the former imperial moat, and beyond to the distant Ming Tombs. In short, the Chen-Liang plan of 1950 was an extraordinary exercise in the blend of traditional and modern, Chinese and foreign design concepts. Perhaps for that reason, however, so too was it a plan never to be implemented.

In fact, the Chen-Liang plan was an early casualty of anti-Western sentiments exacerbated by the Korean War, and more directly by the almost total alignment of the PRC with the Soviet Union and the arrival in Beijing of hundreds of Soviet advisors and technicians during the first nine years of the new regime, especially from 1953 to 1958. Even more pointedly, the Chen-Liang plan and its authors fell victim to the political furor of the Hundred Flowers Campaign and Anti-Rightist Movement (1957) and the Great Leap Forward (1958–60), the most powerful landscape consequence of which was the design, construction and completion of Tiananmen Square in 1959. Simply stated, to commemorate the tenth anniversary of the founding of the PRC, in late 1958 decisions were made by the CCP leadership to rush the construction of a giant 40-square-hectare public square immediately in front of the Gate of Heavenly Peace (*Tiananmen*). Borrowing design and engineering inspiration from Soviet advisors, the square was bordered on the west by a massive 171,800-square-meter Great Hall of the People (the national "parliament" whose main hall was designed to accommodate 10,000 delegates) and on the east by the Museum of Chinese Revolution and History. At the center of the square, a marble obelisk, carved with scenes depicting different episodes in the revolution, would serve as the Monument to the People's Heroes. And all of this architectural space was constructed within months and completed by 1 October, 1959 to stand as the symbolic and operational core of China and of Beijing. Not coincidentally, it stood as well in accordance with arcane prescriptions exactly on the once sacred axis of Celestial Beijing. It was, in sum, a bold statement in the powerful continuities of China.

To be sure, Tiananmen Square and its component buildings were only barely Chinese in appearance. Designed to suit the neo-classical (Greco-Roman) proclivities of Soviet planners, the square and its buildings lacked all but the most incidental of traditional Chinese architectural treatments. Moreover, the giant square was not the taboo-space of Imperial China but the public-space of Socialist China, a place intended to be filled with a popular and plebian presence whose spatial mass was commensurate with the proletarian ambitions of socialism (Hou Renzhi & Wu Liangyong 1977). Indeed, now opened at its northern end by the expansion of Changan Avenue into a major arterial, the new core also served to shift the traditional north–south axis of Imperial Beijing toward a new, "Socialist east–west axis" (Hou Renzhi 1973, 1984, 1986).

Nevertheless, in making use of the traditional cosmology, Tiananmen Square became a powerful statement in the many ambiguities of modern China. A new, revolutionary and socialist China was paradoxically founded upon the arcane, cosmo-magical hub of Celestial Beijing. Indeed, the paradox would find even greater expression 17 years later when, in 1976, Tiananmen Square became the site for the Mausoleum of Mao Zedong. That it was the extreme left-wing iconoclasts of the CCP who insisted on this site for the interment of the remains of their once "Great Helmsman," may be taken as a hardened measure of the paradoxical and lasting geomantic powers of the traditional axis of Beijing. It was paradoxical, moreover, not only because it rested on the traditional axis, but also because the choice of that site broke with all previous precedent. After all, no emperor was ever laid to rest anywhere in the Celestial City. Chairman Mao was the first and only ruler of China, and probably the only person, ever to be interred anywhere near the sacred axis – an act that symbolized his unique singularity, as well as his virtual deification. Furthermore, if a design was chosen for the mausoleum that would fit more comfortably in Athens, Rome or New York than in the capital of China, it was nonetheless in keeping with the overall architectural ambiguity and the historical paradox of Tiananmen Square.

Whatever else they may have intended, in short, in Tiananmen Square the leadership of Socialist China both restored and transformed the symbol and the magic of Celestial Beijing. Not coincidentally, one of the casualties of that restoration was the Chen-Liang plan. Indeed, they not only rejected the Chen-Liang plan, but also condemned the authors of that plan as "rightists" to be purged from the body politic. Ironically, for having sought to move the symbolic core of Socialist China away from the symbolic core of Imperial China, Chen and Liang were labeled "rightists" and reactionaries," while those who restored the arcane symbolism of the traditional core were entitled "progressives" and "revolutionaries." Such, no doubt, is the stuff of Orwellian double-think, but such too is the paradoxical salience of the past in modern China and, in particular, on the landscape of modern Beijing.

Conclusion: the power of place

Nowhere else in the landscape of modern Beijing would the paradoxical salience of the past be more powerful or more obvious than in the case of Tiananmen Square. However, the ambiguities in the search for such salience can also be witnessed in continuing debates among architects, planners, policy makers and the general public over questions of design authenticity and historic preservation. Indeed, despite the iconoclastic rampage of the Cultural Revolution, what had already been in the 1920s and 1930s, and what had promised to become again in the early 1950s a "Chinese Renaissance" has, in fact, reemerged in the 1980s to once again challenge, question and derive strength from the heritage of the past.

A full-scale, national preservationist movement has emerged in China since 1980 (Olpadwala & Tomlan 1985, Samuels, M. & C. 1986). Virtually every county, city and province, along with the national government, has promulgated all manner of new laws and regulations aimed to protect the landscape endowment of Imperial China (Li Zhun 1984, Zhao Xiqing 1984, Zhao Xun 1984). At the national level, in November 1982 the Standing Committee of the National People's Congress passed a new "Law of the People's Republic of China for the Protection of Historic and Cultural Relics." The substance and language of the latter was based partly on the series of "Provisional Regulations" promulgated 20 years earlier in 1961, but was mainly derived from a set of "Regulations for the Protection of Historic and Cultural Relics" that had just been issued by the Beijing municipal government in 1981.

As the national capital, Beijing was perhaps only naturally at the forefront of these endeavors. In 1981 two new municipal advisory and administrative units were established to oversee the selection and protection of historic sites. From 1982 to 1983 the city designated 78 sites for protection, including 63 major traditional sites (e.g., the Palace Museum, Temple of Heaven, Ming Tombs, etc.) and 15 so-called revolutionary historic sites (e.g., the houses of Song Qingling, Guo Morou, and Luxun, Lukouqiao or Marco Polo Bridge, etc.) (Zhao Xun 1984). Furthermore, in July 1983 a new "Master Plan for the Development of Beijing" was approved by the State Council and the Central Committee of the CCP, no small part of which included a series of regulations to protect historic sites and to preserve the "character" of the city. And, by late 1985 the list of protected historic sites and buildings had expanded to include 189 sites under the direct management of the municipal government, as well as some 459 sites under the management of local district and county authorities

(Zhu Changling 1985, 1986). By way of comparison, at the end of 1985 there were throughout China only 242 historic sites under the direct management of central government agencies, and about 4,000 such sites under the management of provincial government agencies (*China Daily*, 25 June, 1986, p. 1).

We need not here detail further the recent record of preservationist efforts in China or in Beijing in order to underscore the density of concerns for the landscape endowment of Confucian China. Should additional quantitative evidence be required, we need only note that the Ministry of Culture and its "Culture Relics Protection Research Department" in 1981 began what was to have been a five-year national survey of historic sites, one by-product of which was to have been a national atlas and registry. According to recent press releases, by June 1986 the survey had already covered more than 100,000 such sites (*China Daily*, 25 June, 1986, p. 1).

Still, it is not so much the number of protected sites or the quantity of places preserved and restored that underscores the power of particular places to evoke the past, or the many dilemmas of cultural salience in modern China. Rather, it is the extraordinary durability of the search itself – of the struggle for design authenticity and historic preservation amidst all of the revolutionary transformations of the 20th century – that underscores the paradox of a China now modern yet still traditional. In microcosm, furthermore, it is a search and a paradox the results of which are nowhere more powerfully articulated than in the triumphs and tribulations of the city that is at once quintessentially Chinese and Socialist, Beijing. Indeed, the paradox of modern Beijing was perhaps best summarized by Chen Xitong, the Mayor of Beijing, when, in addressing a conference of planners and architects in April 1986, he noted that, "Beijing cannot in good conscious honor its ancestors or face its descendants if the ancient beauty and heritage of the Chinese capital is not preserved" (*China Daily*, 28 April, 1986, p. 3). That the Mayor of Socialist Beijing, a senior member of the Chinese Communist Party, could make so blatantly Confucian a comment is perhaps only to say that Beijing has refused to give up its powers as the celestial core of China even as it seeks a revolutionary future. But, it is also to say that Beijing – like China as a whole – remains caught somewhere in the margins between past and future, and that its powers as a place are finally those that epitomize the "central Problematik of contemporary China," the battle for cultural integrity.

References

Aisin-Gioro Pu Jie 1985. *Pu Yi likai Zijincheng yihou.* (After Pu Yi Left the Forbidden City). Beijing: Wenshi Cailiao Chubanshe.
Aisin-Gioro Pu Yi 1964. *From Emperor to Citizen – The Autobiography of Aisin-Gioro Pu Yi.* Translated by W. T. F. Jennes. Vol. 1. Peking: Foreign Language Press.
Bai Rixin 1983. Yuanming Zhangchun Qichun san yuan xingxiang de tantao. (An enquiry into the forms of the three gardens: Yuanming, Zhangchun and Qi Chun). *Yuan Ming Yuan* 1, pp. 22–31.

Bredon, J. 1982. *Peking: a historical and intimate description of its chief places of interest.* Shanghai: Kelly & Walsh, Ltd, 1931. Reprinted in Hong Kong: Oxford University Press.

Butterfield, F. 1987. *China: alive in a bitter sea.* New York: Times Books.

Chang Sendou 1970. Some observations on the morphology of Chinese walled cities. *Annals of the Association of American Geographers* **60**: pp. 63–91.

Chang Sendou 1977. The morphology of walled capitals. In *The city in late Imperial China,* G. W. Skinner, (ed.) pp. 75–100. Stanford: Stanford University Press.

Chen Zhanxiang & Liang Sicheng. 1983. Beijing guihuatu jiani fangan (Proposed map for the plan of Beijing, 1950), partly reprinted in *Chengshi Guihua.* 6, inside back cover. Also see Liang Sicheng, 1986, below.

Cody, J. 1986. An American architect's 'renaissance' in China: Henry K. Murphey's First Decade, 1914–1923. Unpublished paper. Cornell University, School of City and Regional Planning, and Department of History.

Cohen, P. 1984 *Discovering history in China.* New York: Columbia University Press.

Du Xianzhou, (ed.) 1983. *Zhonqquo qujianzhu xiushan jishu* (Techniques for the Restoration of Ancient Chinese Architecture). Beijing: Ministry of Culture, Cultural Relics Protection Research Department, Zhongguo jianzhu gongye chubanshe.

Dunfield, D. M. & P. J. Graham. 1986. Conservation and integration of historic architecture in contemporary Japan. *Orientations* (June), pp. 42–53.

Gamble, S. & J. S. Burgess. 1921. *Peking. a social survey.* New York: G. H. Doran Company.

Han Ji 1984. Preservation of the historic city of Xian. *Building in China, Selected Papers* **2**, pp. 9–15. Beijing: China Building Technology Development Centre.

Hong Yeh (William Hong) 1981. *Hong Yeh Lunxueji* (Collected Writings of Hong Yeh). Beijing: Zhonghua Shuzhu.

Hou Renzhi 1973. Beijing jiucheng pingmian sheji de gaizao (Changes in the design of the old city of Beijing). *Wenwu* **5**: pp. 2–13. Reprinted in *Lishidilixue lilun yu shijian* (see below pp. 205-26.)

Hou Renzhi & Wu Liangyong. 1977. Tiananmen guangchang lizan cong gongting guangchang dao renmin guangchang de yanbian he gaizao. (The transformation of Tiananmen Square from palace square to people's square). *Wenwu* **9**, 1–15. Reprinted and edited in *Lishidilixue lilun yu shijian* (see below, pp. 227–250.)

Hou Renzhi. 1984a. *Lishidilixue lilun yu shijian* (The theory and practice of historical geography). Shanghai: Renmin chubanshe.

Hou Renzhi. 1984b. Historical geography in China. *Geography in China* Beijing: The Geographical Society of China, Science Press. pp. 133–46.

Hou Renzhi. 1984c. Beijing: historical development and reconstruction. *Building in China, Selected Papers* **2**, pp. 1–5. Beijing: China Building Technology Development Centre.

Hou Renzhi. 1985. Lun Beijing jiucheng de gaizao. (Discussion on the changes in the old city of Beijing). In *Chengshifazhan zhanlue yenjiu*: pp. 219–37. Beijing: Xinhua chubanshe.

Hou Renzhi. 1986. Evolution of the city plan of Beijing. *Third World Planning Review* **8**: pp. 5–17.

Hsu, Immanuel C. Y. 1970. *The rise of modern China.* New York: Oxford University Press.

Jiudu wenwulue (Outline of the Cultural Relics of the Old Capital). 1935. Beijing: Beijing shizhengfu mishu zhu.

Johnston, R. F. 1934. Peking. *The Geographical Magazine* **1**, pp. 185–97.

Johnston, R. F. 1985. *Twilight in the Forbidden City*. Hong Kong: Oxford University Press. First published by V. Gollancz Ltd, 1934.

Kwok, Yinwang. 1981. Trends of urban planning and development in China. In *Urban development in modern China*. Lawrence J. C. Ma, (ed.), pp.147–93. Boulder, Colorado: Westview Press.

Levenson, J. R. 1965. *Confucian China and its modern fate*. Berkeley: University of California Press.

Leys, Simon 1976. *Chinese Shadows*. New York: Viking Press.

Leys, Simon 1986. *Burning forest: essays on Chinese culture and politics*. New York: Holt, Rinhehart & Winston.

Li Xiongfei 1985. *Chengshi guihua yu gujianzhu baohu*. (Urban planning and the preservation of ancient buildings). Tianjin: Tianjin kexue jishu chubanshe.

Li Zhun. 1984. Urban planning and historic preservation in Beijing. *Building in China, Selected Papers* 2, pp. 6–8. Beijing: China Building Technology Development Centre.

Liang Quichao 1903. Niuyueh bowuyuan chenlie Yuanmingyuan pinwu jian zhi hanyan. (Shame Upon Viewing Objects from the Yuanming Yuan on display in the Metropolitan Museum of New York). Reprinted in *Yuan Ming Yuan* 1983. **2**, p. 17.

Liang Sicheng (also Liang Ssu-ch'eng) 1983. Song *Yingzao Fashi* zhushi xu. (Preface to Annotations on the song dynasty text, *Yingzao Fashi*). *Jianzhu shilun wenji* **1**, pp. 1–9. (Original manuscript, 1963).

Liang Sicheng & Chen Zhanxiang 1983. Beijing guihuatu jianyi fangan (Proposed Map for the Plan of Beijing, 1950). Partly reprinted in *Chengshi Guihua* **6**, inside back cover. Also see Liang Sicheng, 1986, below.

Liang Sicheng 1984. *A pictorial history of Chinese architecture. A study of the development of its structural system and the evolution of its types*, W. Fairbank (ed.). Cambridge, MA.: The MIT Press.

Liang Sicheng & Chen Zhanxiang. 1986. Guanyu zhongyang renmin zhengfu xingzheng zhongxinchu weizhi de jianyi. (Regarding a proposal for the site of the administrative core of the central people's government). Original February 1950. *Liang Sicheng wenzhi* **4**, pp. 1–31.

Liu Dunzhen 1957. *Chung-kuo zhuzhai kaishuo* (General description of Chinese residences). Peking: Jianzhu gongcheng chubanshe.

Liu Dunzhen. 1981. Tongzhi chongxiu Yuanmingyuan Shiliao (Historical materials on the reconstruction of the Yuanmingyuan during the reign of the Tongzhi Emperor). *Yuan Ming Yuan* **1**, pp. 121–71.

Lowenthal, D. 1985. *The past is a foreign country*. Cambridge: Cambridge University Press.

Lu Xun 1980. *Selected works of Lu Xun*. Beijing: Foreign Language Press. 4 vols.

Luo Ge 1986. Jieguan Beiping gugong bowuyuan suoji (A vignette on the takeover of the Beijing Palace Museum). *Yen Du* **2**, p. 23, 43.

Luo Zhewen 1981. Wei shenma yao baohu gujianzhu (Why ancient buildings ought to be preserved). *Jianzhu lishi yu lilun* (Corpus of architectural history and theory) Nanjing: Jiangsu People's Press and The Sub-Committee on Architectural History of the Architectural Society of China. **1**, pp. 30–5.

Ma, L. J. C. 1981. Urban housing supply in the People's Republic of China. In *Urban development in modern China*, L. J. C. Ma. (ed.), pp. 222–59. Boulder, Colorado: Westview Press.

Mao Zedong 1959. *Talks delivered at the Yenan Forum on artistic and literary work*, Beijing: Foreign Language Press.

Malone, C. B. 1929. Current regulations for building and furnishing Chinese imperial palaces, 1727–1750. *Journal American Oriental Society* **4**, pp. 19–38.

Malone, C. B. 1934. History of the Peking summer palaces under the Ch'ing dynasty. *Illinois Studies in the Social Sciences* **XIX**. Urbana, Illinois: University of Illinois.

Norberg-Schulz, C. 1975. *Meaning in western architecture*. New York: Praeger.

Olpadwala, P. & M. A. Tomlan. 1985. *Summary Report of the Cornell University Delegation on Historic/Cultural Preservation to the People's Republic of China. Beijing: August 12–20, 1985*. Ithaca: Program on International Studies in Planning and Department of City and Regional Planning, Cornell University.

Rapoport, A. 1969. *House form and culture*. Englewood Cliffs: Prentice- Hall.

Rapoport, A. 1982 *The meaning of the built environment*. Beverly Hills: Sage.

Samuels, M. S. 1979. The biography of landscape. In *The interpretation of ordinary landscapes*, D. W. Meinig. (ed.), pp. 51–88. New York: Oxford University Press.

Samuels, C. M. 1986. Cultural ideology and the landscape of Confucian China: the traditional Si He Yuan. M. A. Thesis. Vancouver, B.C.: University of British Columbia, Geography Department.

Samuels, M. S. & C. M. Samuels 1986. The politics of historic preservation in modern China: 1860-1985. Paper prepared for the International Geographical Union, Regional Conference, History of Geographic Thought Committee, Barcelona, Spain. 26–29 August.

Shen Yulin & Li Xiongfei 1985. Lishi wenhua ming (gu) cheng de guihua jiegou yu tese (The structure and special features of the planning of ancient historical and cultural cities). *Chengshi fazhan zhanlue yenjiu*. Beijing: Xinhua chubanshe. pp. 238–53.

Siren, O. 1924. *The walls and gates of Peking: researches and impressions*. London: John Lane.

Siren, O. 1926. *The imperial palaces of Peking*. 3 vols. Paris.

Siren, O. 1949. *Gardens of China*. New York: The Ronald Press.

Spence, J. D. 1981. *The gate of heavenly peace*. New York: Viking Press.

Su Gin-djih 1964. *Chinese architecture: past and contemporary*. Hong Kong: The Sin Poh Amalgated (HK) Limited.

Tong Xi 1981. Beijing Changchunyuan xiyang jianzhu. (Western architecture in the Garden of Lasting Spring of the Yuanmingyuan). *Yuan Ming Yuan* **1**, pp. 71–80.

Tu Weiming 1979. Confucianism: symbol and substance in recent times. In *Value change in Chinese society, R. W. Wilson, et al.* (eds). New York: Praeger.

Wakeman, F. Jr 1973. *History and will: philosophical perspectives of Mao Tse-tung's thought*. Berkeley: University of California Press.

Wang Canchi (ed.) 1985. *Beijing shidi fengwu shulu*. (Annotated bibliography on the history, geography and cultural sites of Beijing). Beijing: Beijingshi shehui kexue yenjiusuo and Beijing chubanshe.

Wang Zhili 1981. Yuxiao baohu Yuanmingyuan yizhi yu jiji kaichan kexue yenjiu. (Effective preservation and intensive scientific research on the relic Yuanmingyuan). *Yuan Ming Yuan* **1**, pp. 16–20.

Wang Zhili 1981. Fengjing chukou qiantan (Elementary comments on foreign tourism). *Jianzhu lishi yu lilun wenji* **1**, p. 10.

Wang Zhili 1983. Yuanmingyuan yizhi shengxiu chutan. (Preliminary explorations on the reconstruction of the Yuanmingyuan). *Yuan Ming Yuan* **2**, pp. 5–14.

Wang Zhishu, Liu Wei & Wang Chaohui 1983. *Beijing Fandian shiwen* (Vignettes on the history of the Beijing Hotel) Beijing: Gongren chubanshe.

Wheatley, P. 1971. *The pivot of four quarters*. Chicago: Aldine.

Wright, A. F. 1977. The cosmology of the Chinese city. In *The city in late Imperial China*, J. W. Skinner. (ed.) pp. 33-73. Stanford: Stanford University Press.

Wright, M. C. 1965. *The last stand of Chinese conservatism: the T'ung-chih restoration, 1862–1874*. New York: Atheneum.

Wu Guangzu 1983. Shilun wo guo jianzhu xinfengge de chuangzao jingyan. (On the creation of new architectural styles in China). *Jianzhu shilun wenji* (Treatises on the history of architecture). Beijing: Qinghua University Press, *Journal of the School of Architecture, Qinghua University* 1, pp. 96–111.

Wu Liangyong 1979. Beijingshi guihua chuyi. (Some comments concerning the city plan of Beijing). *Jianzhu shilun wenji* (Treatises on the History of Architecture). Beijing: Qinghua University Press, *Journal of the School of Architecture, Qinghua University* 3, pp. 167–76.

Wu Liangyong 1983. Lishi wenhua mingcheng de guihua jiegou, jiucheng gengxin yu chengshi sheji. (Planning structure of famous ancient historical and cultural cities, renewal of old city cores and urban design). *Chengshi guihua* 6, pp. 2–12.

Wu Liangyong 1986. A brief history of chinese city planning. *Urbs et Regio* 38. Kassel: Kasseler Schriften zur Geographie und Planning.

Xinxinxiangrong de Beijing (Flourishing Beijing) 1984. Beijing: Beijingshi Tongjizhu (Beijing Municipal Statistics Department). Beijing chubanshe.

Xu Zhaokui 1981. Yu Gong xuehui zhi lishidili yenjiu. Research on historical geography of the Yu Gong Society). *Lishi Dili* 1, pp. 211–20.

Yee, Francis 1984. An historical geography of book markets in traditional China: the case of Luilichang. M.A. thesis. Vancouver, BC: Department of Geography, University of British Columbia.

Zhang Zugang 1981. Cong Zuozhengyuan shili tan Suzhou gudian yuanlin di guwei jinyong. (On the problem of making ancient Suzhou gardens serve the present in light of the example of the Zhuozheng Garden). *Jianzhu lishi yu lilun* (Corpus of Architectural History and Theory). Nanjing: *Jiangsu People's Press and The Sub-Committee on Architectural History of the Architectural Society of China* 1, pp. 52–65.

Zhang Jingxian 1981. Luxun lun jian-zhu (Luxun on architecture). *Jianzhu shilun wenji* (Treatises on the History of Architecture). Beijing: Quinghua University Press, 1981; (*Journal of the School of Architecture, Qinghua University*) 5, pp. 107–18.

Zhang Kaiji 1985a. Beijing: An old cultural capital in change. Unpublished paper for Cornell University seminar on Historic/Cultural Preservation held at Beijing University, 12–20 August. See Olpadwala above.

Zhang Kaiji 1985b. Yao aihu Beijing cheng. (Lovingly protect the city of Beijing). *Yen Du.* 2: p. 4, 6.

Zhang Kaiji 1986a. Qieting waiguo baijia yen: Zaitan Beijing gaoceng jianzhu wenti. *Yen Du* 3, pp. 10–11.

Zhang Kaiji 1986b. Weihu Beijing fengmao, fayang zhongguo wenhua. (Protect the style of Beijing, develop the culture of China). *Fayen* (Speech) 1–6. Samuels collection.

Zhang Kaiji 1986c. Yaoi baohu Beijing lishi wenhua mingcheng de mianmao. (Preserve the famous historical and cultural face of Beijing). *fayen* (Speech). 4–6. Samuels collection.

Zhao Guanghua 1984. Changchunyuan jianzhu ji yuanlin huamu zhi yixie ziliao. (Some materials on the buildings and garden plantings of the Changchun Garden). *Yuan Ming Yuan* 3, pp. 1–11.

Zhao Xiqing 1984. Woguo chengshi guihua gongzuo sanshi nian jianji (1949-1982). (A brief review of thirty years of urban planning in China). *Chengshi guihua* (City Planning Review). 1, pp. 42–8.

Zhao Xun 1984. 'Zhonghua Renmin Gongheguo wenwu baohu fa' zhong you guan
gujian yuanlin shiyong guanli he weixiu fangmian de guiding' (Regulations on
those aspects of the 'Law of the People's Republic of China for the Protection
of Cultural Relics' concerning the Use, management and repair of ancient
buildings and gardens). *Gujian yuanlin jishu* 2, pp. 51–2

Zheng Ruolin, (Trans.) 1984. Yuguo nuchi dangnian Ying-Fa quinluezhun jielue
fenshao Yuanmingyuan de zuixing. (Victor Hugo's contemporary condemnation
of the crime committed by the combined British-French army in looting and
burning the Yuanmingyuan-1861). *Yuan Ming Yuan* 3, pp. 135–6.

Zhou Weiquan 1985. Yuanmingyuan. In *Yuanmingyuan, Zhonqquo lishi shang di
yidai mingyan* (The Yuan Ming Garden, famous historical garden of China),
pp. 64–83. Hong Kong: Joint Publishing Co.

Zhu Maowei 1984. The master plan of Yangzhou, A city of historic and cultural
tradition. *Building in China, Selected Papers* 2, pp. 16–21. Beijing: China Building
Technology Development Centre.

Zhu Changling (also Ju Changling) 1985. Administration and management of
orders of preservation of major architectural monuments in Beijing. Beijing
Administrative Bureau for Museums and Archaeological Data. Unpublished
paper for Cornell University seminar on Historical/ Cultural Preservation held at
Beijing University, 12–20 August. See Olpadwala above. A Chinese language
version may be found in the following item.

Zhu Changling 1986. Yiding yao baohu Beijing de gudu fengmao. (We must
protect the style of the old capital of Beijing). *Yen Du* 1, pp. 4–6.

Zong Quanchao 1985. Wangfujing dajie de xingqi he fazhan. (The rise and
development of Wangfujing Avenue). *Beijing Shiyuan* 3, pp. 128–37. Beijing:
Beijingshi shehui kexue yenjiu suo and Bejing chubanshe.

Index

229

Milton Keynes UK
Ingram Content Group UK Ltd.
UKHW031147141024
449569UK00024B/993